ASSESSING
STUDENT
UNDERSTANDING
IN SCIENCE

Sandra K. ENGER · Robert E. YAGER

ASSESSING
STUDENT
UNDERSTANDING
IN SCIENCE

A STANDARDS-BASED
K-12 HANDBOOK

CORWIN PRESS, INC.
A Sage Publications Company
Thousand Oaks, California

Copyright © 2001 by Corwin Press, Inc.

For information:

Corwin Press, Inc.
A Sage Publications Company
2455 Teller Road
Thousand Oaks, California 91320
E-mail: order@corwinpress.com

Sage Publications Ltd.
6 Bonhill Street
London EC2A 4PU
United Kingdom

Sage Publications India Pvt. Ltd.
M-32 Market
Greater Kailash I
New Delhi 110 048 India

Printed in the United States of America

Library of Congress Cataloging-in-Publication Data

Enger, Sandra K.
Assessing student understanding in science: A standards-based K-12 handbook / by Sandra K. Enger & Robert E. Yager.
p.   cm.
Includes bibliographical references and index.
ISBN 0-7619-7648-5 (cloth: alk. paper)
ISBN 0-7619-7649-3 (pbk.: alk. paper)
1. Science—Study and teaching (Elementary)—Standards—United States
2. Science—Study and teaching (Secondary)—Standards—United States.
3. Science—Ability testing—United States—Handbooks, manuals, etc.
I. Yager, Robert Eugene, 1930– II. Title.
LB1585.3 .E53   2000
507.1—dc21   00-009205

This book is printed on acid-free paper.

01 02 03 04 05 06 07 7 6 5 4 3 2 1

Corwin Editorial Assistant: Catherine Kantor
Production Editor: Denise Santoyo
Editorial Assistant: Cindy J. Bear
Designer/Typesetter: Barbara Burkholder/Lynn Miyata
Indexer: Kathy Paparchontis

Contents

Preface

Educators have responsibilities for not only setting educational goals and objectives but also for instructing and assessing in ways that help students attain these goals and objectives. Assessment has become an increasingly important part of teachers' professional practice. Teachers make decisions about how to interact with their students at the rate of 1 out of every 2 to 3 minutes, and they base those decisions on their own assessment of student learning (Stiggins, 1994). Because of the relationship of assessment and the instructional process, assessment has continued to be a focus in education today.

Assessment, when integrated with instruction, can provide a basis for restructuring science education, and the National Science Education Standards (NSES; National Research Council [NRC], 1996) are incorporated in *Assessing Student Understanding in Science: A Standards-Based K-12 Handbook.* In the past, assessment practices have often focused on the use of set questions that have provided a limited number of options for student responses. The current emphasis on assessment at local, state, and national levels means that educators must make changes in assessments that are implemented in classrooms. Assessment can be viewed as a pathway to address the following questions:

- Should assessments tell us what students cannot do or what each student can do?

- Should assessments set goals for learning or merely sample the present curriculum?

- Should students be judged only on their individual work or also on their abilities to work together for the benefit of a larger group?

- How can assessments encourage and recognize inventive, imaginative responses that, although unexpected, are constructive and appropriate?

- To what extent can students evaluate data, understand concepts, demonstrate process mastery, and apply what has been learned to new situations?

- How does one assess that each student can actually do what the instruction intends for him or her to do? What evidence is used to document this assessment?

- What can be done to help students become better learners?

- How can students attain the desired achievement levels?

Effective educational programs are linked to assessment schemes that help students grow, develop, and succeed, and such assessment schemes should be designed to meet the stated instructional goals and objectives of both the teacher and the learner. *Assessing Student Understanding in Science: A Standards-Based K-12 Handbook* provides guidelines that are based on research and examples from educators who have drawn on their work in kindergarten through university settings. *Assessing Student Understanding in Science: A Standards-Based K-12 Handbook* addresses the assessment of student performance and the establishment of criteria on which to base student progress in six domains within science. These domains relate to concepts, processes, applications, attitude, creativity, and the nature of science.

Each domain is described, and a rationale is provided for assessing student learning in that domain. Chapter 1 provides an overview, supported by the research literature, for each of the six domains of science. In Chapter 2, assessment is set in the contexts of teaching, and in this chapter, assessment practices beyond the traditional paper-and-pencil test are addressed. Some examples of assessment alternatives that can be implemented include concept mapping, clinical interviewing, portfolios, videotaping, journaling, brainstorming, open-ended questioning, and a self-report knowledge inventory. Embedding assessment in teaching practice and World Wide Web sites that support assessment practices conclude the chapter.

Evaluating teaching practice provides the context for Chapter 3, and possibilities addressed include action research, videotapes, and journals. Instrumentation to evaluate classroom practice and surveys to assess the alignment of classroom science inquiry learning opportunities with the NSES are located in this chapter. Both student and teacher forms of the science-as-inquiry surveys have been designed so that perceptions can be compared.

Rubrics and scoring guides are the focus of Chapter 4, which describes ideas for designing schemes to assess student work. Some examples of rubrics that have some design problems have been purposely placed in this chapter to highlight some of the difficulties in the design process. Because districts must examine the alignment of their curricula with the NSES, an example of a tool on which to base this process is included.

Chapters 5, 6, 7, and 8 set out assessment examples for multiple grade levels, but often, ideas can be modified for use at various grade levels. In these chapters, the assessment examples address the six domains of science. Chapter 5 has assessment examples that have possibilities for use across all grade levels. Chapter 6 targets Grades K through 4, and Chapter 7 has examples recommended for Grades 5 through 8. Chapter Nine presents ideas for assessment in Grades 9 through 12.

This assessment handbook has a history to which many educators have contributed, and many of the initiatory ideas and samples were generated by Iowa teachers and students in their science classrooms. Many of these ideas and samples have undergone an evolutionary process in the hands of those who have edited these samples and ideas in attempts to clarify the thinking and communication intended by these instruments. The refinement process and the task of improving assessment practices have been guided by the assessment tenets that follow.

Assessment Tenets

Assessment as learning:

- Assessment design is guided by the purpose of the assessment.

- Assessment includes multiple measures that are used to inform instruction.

- Preassessment of student understanding is vital in determining preconceptions and should be completed and documented by the teacher prior to the introduction of each new concept.

- Evidence of student learning is documented throughout the year.

- Assessment provides information about what the student *can* do rather than what the student cannot do.

- Assessment is viewed in terms of the growth in learning of each student.

- Students are assessed both on an individual basis and on their involvement in group work.

- Assessment tasks should be meaningful, challenging, and engaging throughout instructional activities.

- Assessment tasks are set in a real-world context and should have relevancy for the student.

- Central to any assessment scheme are process skills, concepts, attitudes, creativity, understanding the nature of science, and applications to the real world.

- Process skills and cognitive behaviors are assessed throughout learning or instructional activities rather than at the completion of a unit or chapter.

- Responsibility for assessment is shared with the students.

- Assessment is implemented throughout each lesson and includes all student activities that relate to the six domains of science.

- Outcomes of various student assessments, including student interest, direct and drive instruction within learning situations.

Acknowledgments

The contributions of the following reviewers are gratefully acknowledged:

Dr. Fred Bartelheim
Professor, College of Education
University of North Colorado, Greeley, Colorado

Mary Ann Sweet
School Counselor, Tomball Elementary
Tomball, Texas

Diane Holben
Instructor, Saucon Valley High School
Hellertown, Pennsylvania and Minneapolis, Minnesota

Chris Watson
Department of Education, University of Florida
Gainsville, Florida

About the Authors

S andra K. Enger is currently a faculty member at The University of Alabama in Huntsville (UAH) and Coordinator of Science Education in the Institute for Science Education. She teaches methods courses for preservice teachers in elementary science, secondary science and mathematics, and secondary humanities and social sciences. She also supervises field experiences for student teacher interns and teaches a graduate assessment course. As a professional development provider, she is currently involved with Exploring Space: The Classroom Connection, a joint venture between UAH and the United States Space and Rocket Center. She also consults in developing assessment materials and conducting project evaluations. She graduated from Winona State (MN) University with a BS in science and a master's degree in biological sciences. Her PhD in science education was awarded by The University of Iowa. She has had extensive classroom experience at both secondary and university levels.

R obert E. Yager is Professor of Science Education at The University of Iowa where he also earned two graduate degrees. He has directed over 100 National Science Foundation projects and has served as chair for nearly 100 doctoral students. Yager has served as president for seven national professional organizations. He has been involved internationally with special ongoing projects in Korea, Taiwan, Thailand, Indonesia, and various places in Europe. Yager's research interests and teaching are involved with Science, Technology, and Society (STS), especially in terms of it as an instructional reform effort. He is currently president of the National Association of STS.

Assessment Based on Six Domains of Science

When humans use scientific knowledge and technology, global awareness becomes critical for environmental protection. As the American Association for the Advancement of Science (AAAS, 1990) stated in *Science for All Americans,*

> What the future holds in store for individual human beings, the nation, and the world depends largely on the wisdom with which humans use science and technology. But that, in turn, depends on the character, distribution, and effectiveness of the education that people receive. (p. vi)

Accordingly, scientific literacy has become a major goal of science education. Although there is no consensus regarding what kinds of science content are necessary for scientific literacy, a scientifically literate person is believed to be one who appreciates the strengths and limitations of science and who knows how to use scientific knowledge and scientific ways of thinking for living a better life and for making rational social decisions.

Learning that fosters scientific literacy should promote the following things:

- Students' inquiry skills and abilities
- Students' abilities to apply what is learned to new contexts

- Students' content and conceptual understanding

- Students' understanding of the nature of science

As shown in the following figure, an assessment framework for science learning and experiences to promote science literacy can be organized around six domains: I. Concepts, II. Processes, III. Applications, IV. Positive Attitudes, V. Creativity, and VI. the Nature of Science.

FIGURE 1.1.

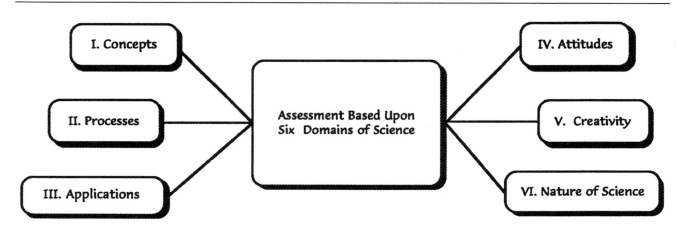

I. Concept Domain

📖 *What Research Says About Science Concepts*

Science concepts are central to science instruction, and student understanding of these concepts is crucial to successful teaching and learning. Millar (1989) noted that without science concepts, it would be nearly impossible for students to follow much of the public discussion of scientific results or public policy issues pertaining to science and technology. According to Thagard (1992), conceptual systems are primarily structured via *kind* or *is-a* hierarchies (i.e., Tweety is a canary, which is a kind of bird, which is a kind of animal, which is a kind of thing) and *part-whole* hierarchies (i.e., a toe is part of a foot, which is part of a leg, which is part of a body). If a basic goal of science education is to help students construct an understanding of the natural world, then students' prior knowledge should be the starting point for instruction.

Assessment enters the field of view to help make the determination of where students are with respect to conceptual understanding. Students should have concrete experience with concepts before moving to abstractions, and they need opportunities to try and to do, not just read about science. The evidence that science concepts have been learned is evidenced most strongly when students can use concepts in a real-life or real-world situation (National Science Teachers Association [NSTA], 1982).

Science in the classroom has been viewed and practiced for decades as a body of knowledge or facts to be learned or absorbed by students. Classically, this occurs by memorization of facts and concepts from a textbook. Knowledge and facts of science are clearly important and indeed necessary, but to memorize these facts as a sole purpose of science education violates the spirit of the very nature of science itself. This issue will be addressed further in the nature of science section.

What the Concept Domain Includes 📖

Facts, laws or principles, theories, and the internalized knowledge held by students fall under the umbrella of the concept domain (Yager & McCormack, 1989). These are the currently accepted scientific constructs related to all of the sciences, and students may best learn these concepts through a curriculum that is conceptually sequenced for developing student understanding. Students must also experience the curriculum from conceptually sound models of assessment and instruction. Science learning should promote conceptual linkages instead of a concepts-in-isolation approach. Concept mastery is an essential aim . . . but only when a meaningful context has been established. Both *Benchmarks for Science Literacy* (AAAS, 1993) and the *National Science Education Standards* (NRC, 1996) provide recommendations about content, concepts, and contexts.

II. Process Domain

What Research Says About Process 📖

Science processes, often designated as *inquiry skills,* are embodied in the terms *exploring* and *investigating.* In science, the investigative processes require hands-on and minds-on activities, laboratory inquiries, and experiments that provide approaches for helping students understand scientific concepts. A study conducted by Shavelson, Baxter, and Pine (1992) found that students with experience with hands-on activities can reliably note their own progress in laboratory activities. More important, these kinds of inquiry skills are also necessary for dealing with everyday life and play a role in the development of an understanding of the natural world (Aikenhead, 1979). The contexts in which the inquiries are set are important in helping students connect the inquiry skills to their personal experiences so that students do not see the processes used in doing science as entities used *only* in science. The application of process skills in a variety of contexts also supports the development of an understanding of the nature of science. Knowledge of the process of constructing and communicating new scientific representations has the potential to yield important insights for science education (Nersessian, 1989).

What the Process Domain Includes

The process domain includes the 13 processes identified by the AAAS (1968) in the development of *Science: A Process Approach*. These are the generally accepted set of processes that scientists use as they accomplish their work. The abilities to use these process skills can be the targets for instruction and assessment, but the identification of separate and distinct processes does not mean that they always occur in definable or identifiable ways. Scientists and students may use several of the science process skills in concert, and these skills may be employed during scientific investigations in ways not expected or predicted by anyone observing the investigative process. These processes and skills are embedded in knowing, doing, and thinking in science.

Process Skills Used in Science

- Observing
- Using space and time relationships
- Classifying, grouping, and organizing
- Using numbers and quantifying
- Measuring
- Communicating
- Inferring
- Predicting
- Identifying and controlling variables
- Interpreting data
- Formulating hypotheses
- Defining operationally
- Experimenting

A Brief Note on How Observation Is Theory Laden

The process of making observations may be influenced by what a person already knows about the subject or object of the observation. Prior knowledge and the conceptual framework that already exist in a person's schema influence the nature and depth of observation. This framework may or not be accurate, but observations are made in the context of this framework. This idea that observation is theory laden may not have been overtly discussed or considered by students. What a person can see depends on what he or she believes, and the observation is therefore theory laden (Abimbola, 1983). Personal viewpoints and creativity also play roles in any investigation, and this brings the implication that the beginning point for investigation should be based on student ideas and questions. Such ideas come from students' prior scientific knowledge, deduction, or even personal guesses and creativity.

A Brief Note About Confirmatory Laboratory Work ✐

Student laboratories and experiments can become exercises in finding the one right answer, and whereas learning a protocol or procedure is often necessary in science, student experiences should move beyond this. Laboratories and investigations should involve the testing of student ideas in which they should have the opportunities to use and develop their abilities with the process domain. Student-generated ideas can serve as the basis for the question or hypothesis that typically precedes any investigation.

A science teacher does need to play a role as the advocate of current "public concepts" (the current accepted scientific thought) to challenge students' "private thought" (Matthews, 1994) or to persuade (Kuhn, 1962) or convince a student to appreciate the current, prevalent interpretation or explanation of natural phenomena. Group discussion for an investigation may produce the same persuading effect (Johnson & Johnson, 1983). Students who understand the role of process skills in scientific investigations may be more likely to see science as a career that is fun and creative.

III. Application Domain

What Research Says About the Application Domain 📖

A key in the application domain is the determination of the extent to which students can transfer and effectively use what they have seemingly learned to a new situation, especially one in their own daily lives (Gronlund, 1988). Students must demonstrate that they not only grasp the meaning of the information and processes but that they can also make applications to concrete situations that are new to them. The application domain is important because it involves having students use concepts and processes not only in a familiar context but in addressing new problems. Students who can apply what they have learned to new situations provide evidence that they have an understanding of a concept.

Two major arenas that students use for applications are in school and in daily life. In school, application often involves problem solving or learning new material by using knowledge and skills acquired in previous studies. In daily life, the crucial factor appears to be the ability to choose the concepts and skills pertinent and relevant for dealing with novel situations. In helping students make applications and connections between science, technology, and their personal lives, the use of current social and technological issues can assist students in seeing the need for the integration of knowledge and skills. Beginning science learning based on students' concerns in the so-called real world may be a way to diminish the learning gap between the world of school science experiences and their personal societal and technological experiences (Yager & McCormack, 1989). An issue-based approach to science learning can serve as a vehicle for engaging students in learning that is local, personal, and relevant.

Dimensions of the Application Domain

- Use of critical thinking

- Use of open-ended questions

- Use of scientific processes in solving problems that occur in daily life

- Abilities to make intradisciplinary connections—integration of the sciences

- Abilities to make interdisciplinary connections—integration of science with other subjects

- Decision making related to personal health, nutrition, and lifestyle based on knowledge of scientific concepts rather than on hearsay or emotions

- Understanding and evaluation of mass media reports on scientific developments

- Application of science concepts and skills to technological problems

- Understanding of scientific and technological principles involved in common technological devices

---

✎ *A Brief Note About Science, Technology, and Society*

Science, Society, and Technology (STS) is the name of an approach characterized by a focus on the integration of science and technology (Yager & Roy, 1993). Local issues provide a science context that has greater student relevancy, and the students learn concepts through activities and community action. In the STS approach, students become involved in resolving local issues and proposing solutions, and this approach also pushes students to seek current information from a variety of sources and experts in various fields of study.

IV. Attitude Domain

What Research Says About the Attitude Domain 📖

How many times have you heard people say that they were never good at science, mathematics, or some other area of study? How important is attitude, anyway? Felker (1974) found that when students were induced to make positive statements about themselves, they attained more positive attitudes about themselves. Page (1958) indicated that teachers who reflected an active and personal interest in their students' progress and who showed it were more likely to be successful in enhancing students' confidence in themselves.

Attitude is very broadly used in discussing issues in science education and is often used in various contexts. Two general categories that are distinguishable are (a) attitude toward science (i.e., interest in science, attitude toward scientists, and attitudes toward social responsibility in science) and (b) scientific attitude (i.e., open-mindedness, honesty, or skepticism; Gardner, 1975). Interest in science tends to decline as students take more science classes and progress through school. This is especially true in the middle school years, and enrollment in science classes declines. Science educators need to work to retain student interest in science and need to consider changing both instruction and assessment practices to be more student centered, to promote ongoing interest.

The positive "I can" attitude and "I enjoy" feelings may enhance students' efforts to seek answers for their own problems and lessen their reliance on others. Students should be able to solve problems with greater independence without parent or teacher intervention. Statements such as, "Don't tell me the answer," or "I can figure it out all by myself," should indicate a growing autonomy. The end result of this self-directed growth could very well be self-acceptance and responsibility for lifelong learning.

The Attitude Domain Calls for Experiences That Support

- Exploration of human emotions
- Expression of personal feelings in constructive ways
- Decision making about personal values
- Decision making about social and environmental issues

- Development of more positive student attitudes toward science in general
- Development of positive attitudes toward oneself (an "I can do it" attitude)
- Development of sensitivity to and respect for the feelings of other people

✎ *A Little Attitude Adjustment, According to Charles Swindoll*

Although this attitude may seem beyond the scope of a science classroom, consider the words of Charles Swindoll (1994):

> The longer I live the more I realize the impact of attitude on life. Attitude to me is more important than facts . . . I am convinced that life is 10% what happens to me and 90% how I react to it. And so it is with you . . . we are in charge of our Attitudes.

V. Creativity Domain

📖 *What Research Says About the Creativity Domain*

Creativity is integral to science and the scientific process and is used in generating problems and hypotheses and in the development of plans of action (Hodson & Reid, 1988). Torrance (1969) defined creativity as the process of becoming sensitive to problems, deficiencies, gaps in knowledge, missing elements, and disharmonies. He also included in his description the identification of the difficulties, the search for solutions, making guesses, or formulating hypotheses about the deficiencies. The testing and retesting of these hypotheses and possibly modifying and retesting them and finally communicating the results all relate to the creative process.

Creativity plays an integral role in the many processes of science and in doing science. Creativity is a complex construct, difficult to assess, and rests very often in what might be called recognizing it when you see it. If a science educator wishes to foster a classroom that enhances students' creativity, this sends a message that the classroom should probably become more student centered. Creativity is fostered and nurtured through a richness in experience. Creativity calls for an openness in the classroom, an acceptance of ideas, a try-new-things approach, and a so-called go-with-the-flow approach. In fact, Csikszentmihalyi (1990, 1996) uses "flow" as descriptive of the state in which creativity is turned on in individuals.

Studies have suggested that the work done in the laboratory rests on the ability to manipulate the objects and the instruments used. Three features of laboratory practice make the need for creative abilities paramount. First, scientists and students do not work with the natural world *as it is* but rather, manipulate the objects of study to make them more accessible for experimentation. Second, investigators do not work with the natural world *where it is* but are instead able to bring those natural objects into an artificial setting (i.e., the laboratory, the classroom, on a slide, etc.). Third, scientists and students do not need to study an event *only when it happens* but rather, can cause the event to occur unnaturally when the situation demands it (Knorr-Cetina, 1981). These three characteristics of a laboratory require an imaginative, inventive mind capable of performing these investigations. These aspects of the scientific enterprise are often ignored in the traditional classroom, yet they are integral to science instruction.

Scientific experiences that can push the creative domain are likely to have some of the following attributes.

The Creative Domain Calls for Experiences That Promote

- Visualization—production of mental images
- Divergent thinking
- Open-ended questioning
- Consideration of alternative viewpoints
- Generation of unusual ideas

- Generation of metaphors
- Novelty—combining objects and ideas in new ways
- Solving problems and puzzles
- Designing devices and machines
- Multiple modes of communicating results

VI. Nature of Science Domain

What Research Says About the 📖 *Nature of Science Domain*

The endeavors undertaken by researchers in their attempts to understand the natural world can serve to promote students' understanding of how science progresses. Science is a human endeavor that relies on reasoning, insight, energy, skill, and creativity (NRC, 1996). Honesty, values, open-mindedness, and in what *Benchmarks for Science Literacy* (AAAS, 1993) denotes as habits of mind, all play roles in scientific ways of knowing. Working with science teachers to develop their understanding of the nature of science is likely necessary prior to their facilitation of instruction that promotes their own students' understanding of this construct. Preservice or inservice courses that emphasize the nature of science can result in significant gains in teacher scores on instruments designed to measure understanding of this construct (Akindehein, 1988; Barufaldi, Bethel, & Lamb, 1977). Selected journal articles, discussions, activities, curriculum projects, and media can be used to help build understanding in this domain.

In the course of human history, people have developed many interconnected and validated ideas about the physical, biological, psychological, and social worlds (AAAS, 1990). Successive generations, enabled by these ideas, have achieved more comprehensive and reliable understanding of the human species and the environment. These ideas have been developed through particular ways of observation, thought, experimentation, and validation. These ways are the bases of what is meant by the nature of science, and they are reflective of how science tends to differ from other ways of knowing (AAAS, 1990).

How scientific knowledge has developed and the roles scientists have played during such a process are two fundamental aspects that are considered important for students to know. Raising student awareness and a development of an understanding of these aspects should be included in science learning. Science itself is dynamic, and as witnessed by history, many ideas have come and have eventually been replaced or discarded. Many science educators suggest that instruction in a science classroom should reflect this tentative nature of scientific knowledge (Lederman, 1992).

That scientific knowledge is tentative has two facets that should be expressed explicitly in working with students. First, the purpose of science is to develop a systematic knowledge to understand how nature behaves. Students should see science as a human endeavor in which scientific knowledge is developed by humans in an attempt to make sense of the world. Accordingly, scientific knowledge is not a truth "to be discovered" in the natural world but a man-made explanation. Second, scientific knowledge can be changed, shifted (Kuhn, 1962) from one point of view to another, due to external social influences, such as politics, economics, and culture. This suggests that scientific knowledge is not absolutely objective. Therefore, the understanding of the involvement of social factors in scientific development provides another purpose for science education.

Science, accompanied by the power of technology, has unique characteristics that affect society. Perhaps no other human activities have ever played such a role in shaping the directions in which societies have moved. The potential to do "good" is very often offset by the power to cause harm, and long-term outcomes and effects are not always predictable.

An important aspect of the nature of science is related to how scientists think and work in the scientific community. Helping students to understand more of the nature of science can promote deeper understanding of what it means to do science. Science is often portrayed as a major intellectual pursuit of truth. From that expectation, many people view scientists as a group of people who are more objective and intelligent than others. Students often believe that scientists can solve problems merely based on their scientific knowledge. Science is a human activity that engages real people.

In doing science, scientists often work collaboratively, and given the specializations in science and related areas, a team approach is very often needed to work on problems. Seldom do scientists work in isolation; a laboratory involves a team of people. In order to ask questions and work at finding solutions to problems, scientists must both share and obtain information from others in their field, and most important, they must reach a consensus by virtue of discussion and persuasion, not just on the basis of mere evidence. Peer review is an important component of doing science, and scientists expect to be challenged and to defend the work they have done. The work that scientists do must be replicable, in that others can verify the work. Science is also a venture characterized by these competitive elements: being first to report findings, competing for research money, achieving status within the scientific community, acquiring status for a university.

Science instruction in the classroom should attempt to portray the nature of the discipline—not simply study the information and interpretation included in the textbook. Views currently held as so-called truths of science have changed and will continue to change throughout time. Therefore, teaching only for the retention of facts without grounding them in real-world experiences will only, sooner or later, result in the loss of these facts from memory. In an attempt to reflect the nature of science, group work, reporting findings, discussion, and reaching consensus are all parameters involved with the nature of science domain.

The Nature of Science Domain Calls for Experiences That Address

- The framing of questions for scientific research
- The methodologies used in scientific research
- The ways in which teams cooperate in scientific research

- The competitive side of scientific research
- The interactions among science, technology, economy, politics, history, sociology, and philosophy
- The history of scientific ideas

Assessment Approaches Aligned With the Six Domains

Assessment approaches should include multiple measures of what students know and can do as a result of their learning experiences. The use of a multifaceted assessment approach has the potential to provide a better profile of student understanding in the six domains, and a more holistic assessment approach deals with the "whole student" (Raizen & Kaser, 1989). Although standardized tests may be valid in measuring knowledge of facts, they may lack validity in measuring higher-level thinking processes, investigation skills, and practical reasoning (Aikenhead, 1973; Champagne & Newell, 1992). This is not to say that standardized tests cannot and do not measure higher-level thinking processes. The deeper issue involved is one of having the assessments align with instruction and intended student outcomes.

Assessment has served as a feedback system to inform teachers about the effectiveness of classroom instruction and to inform students about how well they are learning. Information from assessments has also been used for accountability and policy decisions. The use of information from assessment does influence schools, and teachers may teach to tests knowing that their teaching effectiveness may be judged by their students' scores. Thus, assessment modes can have a profound influence on teaching, often limiting classroom activities to exercises that will be on the tests (Champagne & Newell, 1992). If teachers are compelled to teach to tests, then the tests should emulate the desired student learning outcomes.

The current focus on assessments stresses the necessity to link instruction and learning with assessment. Assessment should be used as a tool for communicating expectations of the science education system to those concerned with science education (NRC, 1996). Accordingly, assessments should be embedded in the learning context to help and guide student understanding on the way toward a cumulative assessment. Having student involvement in the development of assessments and the encouragement of students to self-assess should empower students in taking greater responsibility for their own learning.

The shift from teacher-centered instruction to more student-centered learning may be challenging for both teachers and students. Teachers must adjust teaching practices to include a variety of instructional strategies and should consider teaching and learning environments that move students toward the goal of becoming self-directed learners. The NRC has provided some assessment examples and procedures in the *National Science Education Standards* (NRC, 1996). The standards state that the assessment process is composed of the following components:

- Data use

- Data collection

- Methods to collect data

- Users of data

These four components are interrelated and are used for making decisions and taking action based on the data. The *National Science Education Standards* (NRC, 1996) include the following assessment standards:

Standard A: Assessments must be consistent with the decisions they are intended to inform. (p. 78)

Standard B: Achievement and opportunity to learn science must be assessed. (p. 79)

Standard C: The technical quality of the data collected is well matched to the decisions and actions taken on the basis of their interpretation. (p. 83)

Standard D: Assessment practices must be fair. (p. 85)

Standard E: The inferences made from assessments about student achievement and opportunity to learn must be sound. (p. 86)

2

Assessment in the Contexts of Teaching

Assessment is a process of information collection, and in the contexts of teaching, this information is typically used to examine and describe student performance. The ways in which this information is collected can range from informal to formal, and assessment information can be used in either formative or summative ways or in some combination. The assessment practices used should be linked to student outcomes, and the practices should mirror the ways in which students are learning the information. With any assessment, the underlying purpose for conducting the assessment should guide the assessment design and the use of the results.

Traditional assessments usually fall in the category of the paper-and-pencil type, which are the true-false, fill-in-the-blank, multiple-choice, and short-answer-type tests. When science instruction becomes more student-centered and more constructivist in orientation, a wider range of assessments may be needed to capture student performance. If learning is viewed as an active process (Ausubel, 1968) and if scientific knowledge becomes meaningful to students when they have opportunities to learn science through inquiry (NRC, 1996), then the assessments used should reflect these experiences. With an interactive picture of student learning in mind, current science educators argue that student performance in science should be assessed from various facets of learning in the different contexts of teaching. Performance assessments should be considered for use in an effort to provide multiple measures of student learning outcomes, and performance assessments can more closely parallel the learning experiences. The *National Science Education Standards* (NRC, 1996) have set out assessment standards that can be used to guide assessment practices. The Association for the Education of Teachers in Science is also developing a set of assessment standards that communicate expectations about assessment practices that should be addressed in the preservice science education programs.

The National Council on Measurement in Education (NCME) has also set out the *Code of Professional Responsibilities in Educational Measurement* (NCME, 1995). This code is informational for those involved in various assessment-related practices. Numerous educational measurement books and other resources from organizations such as the regional educational laboratories also provide an excellent source of assessment references ranging from theory to practice.

In this chapter, current assessment types and methods will be reviewed, the outlines of assessments in the *National Science Education Standards* will be introduced, and ways of applying assessment standards in practice will be addressed.

Overview of Assessment Types and Methods

Assessment options are numerous, and those looking at expanding classroom assessment practices will find that many resources are available in print and on-line. The assessment of student performance is likely to include the traditional classroom tests, the commercial standardized tests, and possibly some kinds of performance assessments. These various assessments may be used in concert to provide a picture of what the individual student can do, or in the case of commercial standardized tests, these may be used to provide district-, state-, national-, or international-level information.

Although traditional assessments do provide information, a concern exists that the more traditional standardized tests cannot represent a complete picture of student performance in science (Champagne & Newell, 1992; Pierce & O'Malley, 1992). However, few would deny that standardized tests are indicators of some facets of student achievement, especially in large-scale testing, and should not necessarily be discarded from assessment options. Information from large studies, such as the Third International Mathematics and Science Study (TIMSS) and the National Assessment of Educational Progress (NAEP) can provide assessment information that can at least raise questions about the nature of programs and what students are expected to be able to do.

📖 *Standardized Assessments*

What *Standardized* Means in Assessment

An assessment is standardized in the sense that the administration, scoring, and apparatus are fixed by the assessment developers so that the test may be administered and scored by different examiners in different settings to achieve comparable results across all examinees. A fixed set of test items designed to measure a clearly defined domain and norms characterize standardized tests (Gronlund & Linn, 1990). A standardized test is often administered in a multiple-choice format, and as in the selection of any assessment, various factors play roles in the selection process. For large-scale assessments, a multiple-choice format is often used, and inherent in this format are certain advantages and disadvantages:

Advantages and Disadvantages of Multiple-Choice Tests

Advantages	*Disadvantages*
• Are less costly than some other forms of testing for large-scale testing	• Are expensive to develop and norm
• Are easier to administer to large numbers of students	• Are used to rank schools based on test information
• Make ranking individuals easier	• May be used to judge schools on student performance on one test
• Can be used to make comparisons across locations, cities, states, and nations	• May be used to judge teachers on student performance on one test
• Are less costly than some other forms of testing for large-scale testing	• Can fail to assess higher-order skills
• Can sample a wide variety of learning targets	• Provide just one measure of student learning
• Are fast tools to test student knowledge of specific content	• Are constructed upon the assumption that knowledge can be represented by an accumulation of bits of information and that there is one right answer
• Can assess higher-order skills	
• Can be used diagnostically to improve instruction	• Encourage teaching to the test, which narrows the curriculum

Alternatives in Assessments 📖

What *Alternative* Means in Assessment

Alternative assessments are usually thought of as alternatives to the traditional paper-and-pencil test and are not the commercial standardized multiple-choice test. Alternative assessments tend to be criterion referenced instead of being norm referenced. Alternative assessments can include teacher observations of student performance, student interviews, student self-assessments, presentations, projects, concept maps, portfolios, and any variety of formats used to provide evidence that learning is occurring. Alternatives in assessment options provide a wider range of opportunities for students to represent what they are able to do to meet learning targets.

Alternative assessment enables the evaluation of a wider variety of learning targets, and instructional time can be focused on developing a wider range of student skills and abilities. Furthermore, it is possible to begin to profile a student's intellectual development over time. A major weakness of alternative assessment, at least for large-scale testing, is the difficulty in achieving consistency in interpretation of assessment results (Champagne & Newell, 1992). Alternate assessments also are subject to reliability and validity questions, can be time-consuming to score, and can call for some very subjective judgments.

A recommendation for classroom teachers would be to use multiple measures to assess the cognitive development and academic progress of students. Integration of assessment with instruction, assessment of learning processes and higher-order thinking skills, and a collaborative approach to assessment enable teachers and students to interact in the teaching-learning process.

What *Authentic Assessment* Means

Authentic assessment is viewed by Pierce and O'Malley (1992) as authentic because it is based on activities that represent actual progress toward instructional goals and reflects tasks typical of classrooms and real-life settings. Authentic assessment is also defined as how well the assessment represents a real-world task (Wiggins, 1989a, 1989b, 1993). The most authentic assessment of whether a person can drive a car is to actually drive that car. Wiggins (1998) notes that for authenticity, tasks, problems, or projects should be realistic, require judgment and innovation, have students actually "do" the subject, align with the process in the real-world, possess complexity, and provide opportunities for performance refinement.

✍ *Performance Assessment*

What *Performance* Means in Assessment

Performance assessment is a procedure in which work assignments or tasks are used to obtain information about how well a student has learned (Nitko, 1996). Performance assessments may be alternative assessments, and the actual type and format for the performance may vary. Often, some type of task is involved, and short-term and long-term projects, whether individual or group, could be considered as performance assessments. A wide range of assessments could be subsumed in performance assessment. Performance assessment typically involves setting criteria for which a student demonstrates specific skills and competencies, and a rubric or scoring guide is designed to communicate varying levels of performance. Levels of student performance on the assessment are evaluated on the criteria developed and communicated in the rubric.

Advantages and Disadvantages of Performance Assessment

Advantages	*Disadvantages*
• Provides opportunities for applications of inquiry skills	• Takes time for students to complete
• Provides opportunities to practice with open-ended questions	• Poses tasks that are difficult for some students to complete
• Promotes opportunities for critical thinking	• Calls for subjectivity in assessing performance
• Provides more direct evidence of what students can do versus indirect evidence from traditional tests	• Allows variation in performances and is open-ended in nature
• Promotes opportunities for creativity	• Assesses performance that is task specific
	• Contains time and group management issues
	• Requires time to design, implement, and evaluate student performance
	• Poses difficulty in the design of good tasks and rubrics

✍ *Portfolios in Assessment*

What portfolios are

In the context of assessment, a portfolio is a limited collection of student work that is used to either present best work(s) or to demonstrate a student's educational growth over time (Nitko, 1996). This collection is driven by the purpose for which the portfolio collection is assembled. Portfolios may be used to showcase best work, to provide documentation of long-term projects, and in instances where a purposeful collection provides support and enhancement that cannot be presented in other ways. Pierce and O'Malley (1992) noted that portfolio assessment provided an approach for combining the information from both alternative and standardized assessments, and opportunities for student reflection and self-monitoring were supported by a portfolio approach. As Nitko (1996) so aptly noted, a portfolio is not a scrapbook nor is it a "dumping ground" for all work.

Advantages and Disadvantages of Portfolio Assessment

Advantages	*Disadvantages*
• Provides student practice in decision-making in content selection	• Requires storage of numerous student portfolios
• Promotes student practice in organization	• Requires deciding what kinds of works should be included
• Promotes student practice in communication of work	• Requires deciding how to assess portfolio contents
• Develops student self-assessment strategies	• Requires time to monitor and assess student portfolios
• Provides more opportunity for students to address personal learning styles	• Requires updating portfolios—what to keep and what to weed out
• Provides student opportunities for creative expression	• Requires moving portfolios from one grade level to the next
• Promotes student reflection on personal improvement	
• Promotes student responsibility for maintenance of work history	
• Promotes communication among student, teacher, and parent or guardian	

Portfolio content possibilities: The content that could be included in a portfolio is extensive (see Table 2.1), but decisions about what to include should always be guided by purpose: how the contents will provide support for and evidence of student learning. The portfolio can become a comprehensive form of assessment and documentation of student

TABLE 2.1 Portfolio Content Possibilities

Projects, Notes, Class Work	Various Writing Samples	Journal-Type Entries	Evidence About Self	Media Usage
Individual projects	Self-evaluations	Lab logs	Evidence of effort	Audiotapes
Group projects	Position or issue papers	Multiple formats	Teacher and parent notes	Videotapes
Class notes, research notes, outlines	First and last drafts	Free-write, reflective, prompt-specific comments	Newspaper clippings	Multimedia work
Assignments	Lab reports	Reading response	Comments from peers and others	Photos
Assessments	Creative works—poetry, essays, songs, and so on	Wishes, ideas, thoughts, reactions	Creative endeavors	Artwork
Graphic organizers	Critiques and reviews	"What I understand"	Community involvement	Computer programs
Best works	Letters	"Questions I have"	Areas of greatest improvement	Best works

growth. Student ownership of the portfolio and responsibility for organization should be encouraged.

📖 *Alternative Assessment Implementation Strategies*

In this section, a sampling of some of the many strategies available to help implement alternative assessments are set out. These include observations, clinical interviewing, concept mapping, journaling, brainstorming, open-ended questioning, and self-knowledge reporting.

🖉 *Making and Using Observation*

Teachers are typically observing students in their classes, and these observations are one way to assess student abilities in science. Teachers may not consider the act of observing as a form of assessment, because observations are often filed in a mental notebook rather than being written down. If assessment focuses on the totality of student performance in the classroom, an excellent way to evaluate student work in the laboratory or in the classroom is for the teacher to keep observational records. Direct observation and recording of those observations will give a better

TABLE 2.2 Sample Observation Notation Form

Examples of Student Behaviors	Evidence Observed	Date
Shows curiosity Has a positive attitude Has an investigative spirit Is an independent learner Asks appropriate questions Contributes suggestions Shares group responsibility Contributes to the group Makes responsible use of time Uses lab equipment and facilities with respect Is prepared for class Is receptive to ideas Is receptive to feedback Meets deadlines	*(Notes: The observer could set this up to make notes here. A set of descriptors could be developed. This could also be edited to become a student self-evaluation piece.)*	

account of what students are capable of doing as they work in typical situations. Observational notes can be made, and technology is available to support this, or observation checklists that streamline some of the paperwork can be developed. A generic checklist can be developed with some spaces left for specifics.

Observations can also inform the teacher about which students know how to use the library or media center; how to take notes; how to locate resources; or how to use the computer, microfiche, and the Internet. If assessment is concerned with evaluating the whole student, observation can play a key role in that evaluation, and observations can provide teachers with continued guidance in helping students have maximal success in the classroom.

Observational assessments can provide a more comprehensive picture of student participation and involvement in the classroom. A checklist can be useful for focus and for documentation of observed behaviors. Some examples of student behaviors that could be included as observational evidence are found in Table 2.2.

Time concerns: In life, there is never enough time to do all of the kinds of things that a person wishes to do. Many of the recommendations for changing assessment practice do require time, and if time could be created, the answer for managing assessment may be easier to address. Ask yourself this question: Do the assessment practices in my classroom give the best evidence of what the students know and are able to do? If the answer is yes, you won't be likely to change what is being done. If the answer is no, then perhaps some of the assessment alternatives should be considered and implemented.

✎ Concept Mapping

A concept map is a two-dimensional, graphic representation of understanding of some domain (Novak & Gowin, 1984). Concepts related to the domain of interest are circled and linked to other concepts to show connections. Linking words indicative of the relationship are added to the lines, and a hierarchy is typically present. If the linking words are not included, the basis for linkage must be surmised, and with no linking words, just a web is created. Concept mapping can be used in combination with open-ended questions to provide evidence about what students know and understand. Concept maps provide a graphic look at student understanding, and concept mapping is a skill that must be taught. As an example, the beginning sentences of a response to an open-ended question asking students to describe the water cycle could be used to generate a concept map (see the examples in the accompanying box and Figure 2.1). If most of the words of the sentences were removed, the backbone for a concept map should emerge. Alternatively, students could make concept maps first and then write paragraphs to support their maps.

Written Beginnings for a Concept Map
for the Water Cycle

Water on the surface of the earth is evaporated by the energy of the sun. Clouds form from this water, and this water can eventually fall back to earth as rain or any number of other forms of precipitation.

✎ Clinical Interviewing

A clinical interview follows a protocol for which interview questions have been developed, and these same questions are asked of all persons interviewed. Jean Piaget developed the clinical interview, and it has been adopted by researchers concerned with the prior knowledge, alternative frameworks, or misconceptions held by students and can be used with students after concept mapping (Novak & Gowin, 1984). The interview

FIGURE 2.1. Water Cycle Concept Map Made From Written Beginnings

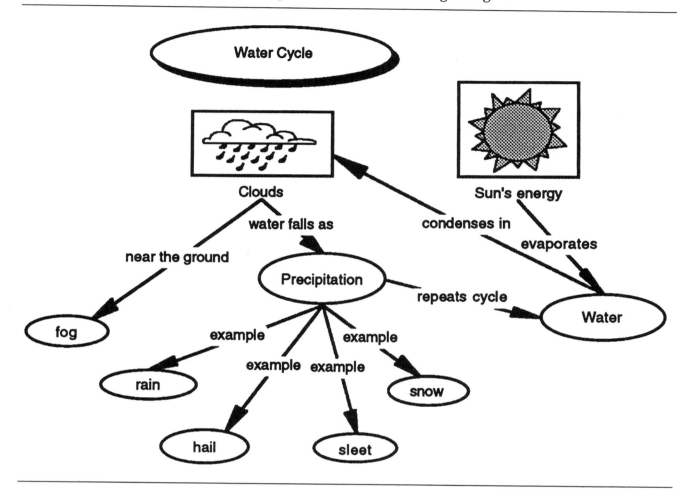

allows the teacher to orally interact with the student and to probe for further clarification of the student's concept map. This can facilitate assessing what the student may be unable to communicate via the written words.

If teachers follow some simple guidelines, the interview can be a powerful tool in establishing students' prior conceptual frameworks and evaluating student learning at any point in instruction. Remember that if the interview is to be audiotaped or videotaped, you may need permission to do so. The time needed to conduct interviews may limit how often this technique is used. However, the ideas underlying the interview are valid for incorporating into questioning strategies for a class.

Here are eight ideas for setting up the clinical interview:

1. Plan the interview. Where possible, use concrete objects, situations, or pictures.

2. Select appropriate questions to ask the student. Ask questions designed only to find out what the student knows about the issue or the concepts involved. Avoid "yes or no" questions. Do not ask questions designed to aid student understanding of the issue. Let the student talk without fear of right or wrong. Have the student describe, predict, or explain.

3. Sequence the questions from the easier to the more difficult ones. It is important to establish student confidence so the student does not become nervous and uncomfortable during the interview. The real goal is to seek explanations.

4. Students should be given adequate time to answer the questions. If the student fails to respond to the initial question or says he or she does not understand the question, the question should be restructured so as to give the student a chance to be successful.

5. Whether it is the first time for teacher or student or the tenth, it is important to convey a relaxed presence. Interviewers should project the image that they are human and that they do not know all the answers.

6. In any sequential interview, it is important to refer to prior interviews or relevant intervening instruction, and the student's ideas should be used. This gives the student a frame of reference from which to respond, makes the conversation "roll," and gives evidence of teacher acceptance

7. The language of the student should be used when rephrasing questions or probing further. To insist on the "right" word or pronunciation can be confusing and inhibit fuller expression of concepts and propositions. Also, students will sometimes use the wrong label (word) for the right concepts. (For example, students will often say that the earth is shaped like a circle rather than a sphere). When this occurs, an explanation can be asked for and the correct concept label supplied by the student.

8. Last, interviews should conclude on a positive note. Cooperation, manners, and answers to questions by the student can be noted; any questions the student might have can be answered. In any case, the interview should end with feelings that will make future interviews an experience to be welcomed and anticipated (Novak & Gowin, 1984).

Concept maps also work very well for group work, and the mapping process engages students in dialogue about the concepts being mapped. Although the resulting concept map is a group product, it is still useful as a preinstruction and postinstruction assessment tool. The concept maps also will vary in organization that provides evidence of alternatives in knowledge representation.

Videotaping ✐

Videotaping can provide an opportunity for self-assessment of teaching practice or assessment of student performance. Audiotaping can also be used with students to help them view their own abilities and knowledge in any particular area. Simply setting up the camera and letting it run during class can provide valuable information for instructional improvement. The tapes can be evaluated with an assessment tool at a later time. If the tapes are to be used for research purposes, then parental or guardian consent is typically required.

Journaling ✐

The journal is a diarylike series of writings or drawings or both. The student should have a separate book or folder for the journal. Entries can be responses to an instructor's questions or statements, feelings about an activity, or "what did I learn today?" The focus on writing is a strength of journals. Students gain practice in their writing skills while communicating their ideas on paper. Journals can take on any variety of formats driven by the intended use of the journal, and creativity can be facilitated with variety in format.

The list of positive aspects of student journals is lengthy. For example, connection of knowledge between subject areas can occur when journals are used. Teachers are better able to respond to individual questions as the journals are read at various times, and a teacher-student dialogue can be established. Long-term improvements within the course and teaching methodology can result in response to insights gained from students.

Brainstorming ✐

Brainstorming can be used as a preassessment strategy to determine what students know about a question or issue they will be investigating. Students could first be asked individually to write down all they know about the topic. After this preliminary session, a class or group brainstorming session could occur. The teacher could at a later point use the papers and notes from the initial session to document and assess growth in that area. When brainstorming serves as an introductory instructional activity, questions are formulated to discover what students know in general that can serve to provide an assessment of what they do not know in the topic area. Brainstorming can also serve as the prelude to concept mapping.

To be certain that every student has a preinstruction assessment, ask each of the students in the class to take a moment to record his or her thoughts, either in a science notebook or on a separate sheet of paper, before the entire-class conversation begins. The exercise of writing before they discuss can also lead to a more engaged class discussion, because the students will have focused on the topic and thought about what they want to say.

At the end of each lesson and at the end of a unit of study, teachers can ask each student to write down what he or she learned. Students can also be asked to write questions that they still have. If the information collected from the different periods of brainstorming is carefully documented and analyzed, it is likely that teachers and students would note growth in understanding from the learning experiences.

✎ Open-Ended Questioning

Open-ended questions invite multiple solutions and multiple ways to arrive at solutions. Open-ended problems assist students in the development of problem-solving skills, as well as promoting creative and divergent thinking. They move students away from trying to guess what the teacher wants and away from the prevalent notion that there is one right answer to all problems. An open-ended assessment approach becomes less of a guessing game and more of a preparation for the real world where multiple solutions are not only possible but often necessary.

Develop open-ended questions by using the following strategies:

- Identify the "big ideas(s)" and subconcepts to be assessed.

- Determine the cognitive skills to be assessed.

- Design a sample prompt (scenario, storyline, etc.) to engage students in thinking and elaboration.

- Write your own sample answer to address this prompt.

- Pilot the open-ended scenarios with students.

- Compare the student responses with your expected outcomes. Rewrite and revise based on student responses.

- Draft a scoring rubric for student responses. Introduce students to rubrics and let them assist with the design and scoring process. When possible share the rubric with students prior to the assessment so students have a context in which to frame a response. As a pool of open-ended questions is developed, save and share exemplary responses with students so students become aware of expectations. Although this means continued assessment development, the assessment itself can become an instructional tool.

Self-Report Knowledge Inventory ✐

For a preassessment of student knowledge in an area to be studied, students can be asked to rate their knowledge of concepts and skills on a five-point scale. Students choose a descriptor of their self-perceived level of understanding in the area of instruction. The instructor sets out some concepts, related ideas, or relevant vocabulary on which instruction will be centered. This may also help to activate prior knowledge. A lesson on ecosystems might focus on trophic levels, and the words set out in the following box might be used.

Sample Self-Report Knowledge Inventory: What I Think I Know About Ecosystems

Use the following statements to indicate what you know about the concepts or words related to ecosystems.

1. I have never heard of this.

2. I have heard of this but do not understand what this is.

3. I think I somewhat understand what this is.

4. I know and understand what this is.

5. I know and understand this well enough to explain this to a another student.

_____ trophic level   _____ food chain

_____ producer   _____ food web

_____ consumer   _____ food pyramid

_____ producer   _____ biomass

_____ decomposer

Teachers can use student responses to direct or guide instruction so students can reach level four or five. This could be given as a preassessment to activate prior knowledge or as a background knowledge probe (Angelo & Cross, 1993). This kind of assessment could also be included as a kind of journal entry, and students could note their changes in understanding as instruction proceeds. The ability of students to self-evaluate is an important goal of education. The time needed to design, administer, and assess with this strategy is minimal, and the results have been found to be valid and reliable (Tamir & Amir, 1981).

National Science Education Assessment Standards

The assessment standards provide criteria to judge the quality of the assessment practices used by teachers, state, and federal agencies to measure student achievement and the opportunity provided for students to learn science (NRC, 1996).

🖉 *Standard A: Coordination with Intended Purposes*

Assessments are consistent with the decisions they are intended to inform. What is expected, to meet this standard?

- Assessments are deliberately designed.

- Assessments have explicitly stated purposes.

- The relationship between the data and the decisions should be clear.

- Assessment procedures need to be internally consistent.

🖉 *Standard B: Measuring Student Achievement and Opportunity to Learn*

Achievement and opportunity to learn science must both be assessed.

- Achievement data focus on the science content that is most important for students to learn.

- Opportunity-to-learn data focus on the most powerful indicators of the students' opportunity to learn.

- Equal attention must be given to the assessment of opportunity to learn and to the assessment of student achievement.

🖉 *Standard C: Matching Technical Quality of Data With Consequences*

The technical quality of the data collected is well matched to the decisions made and the actions taken on the basis of data interpretation.

- The feature that is claimed to be measured is actually measured.

- Assessment tasks are authentic.

- An individual student's performance is comparable on two or more tasks that purportedly measure the same aspect of student achievement.

- Students have adequate opportunity to demonstrate their achievements.

- Assessment tasks and methods of task presentation provide data that are sufficiently stable to lead to the same decisions if used at different times.

Standard D: Avoiding Bias 🖉

Assessment practices must be fair.

- Assessment tasks should be reviewed for the presence of stereotypes, for assumptions that reflect the perspectives or experiences of a particular group, for language that might be offensive to a particular group, and for other features that might distract students from the intended task.

- Large-scale assessments must use statistical techniques to identify potential bias that could contribute to differential performance.

- Assessment tasks must be appropriately modified to accommodate the needs of students with physical disabilities, learning disabilities, or limited English proficiency.

- Assessment tasks must be set in a variety of contexts and engage students with varied interests and experiences, and tasks should not assume the perspective or experience of a particular gender, racial, or ethnic group.

Standard E: Making Sound Inferences 🖉

The inferences made from assessments about student achievement and opportunity to learn must be sound.

- When making inferences from assessment data about student achievement and opportunity to learn science, explicit reference needs to be made to the assumptions on which the inferences are based.

Embedding Assessment Standards in Teaching Practice

In this section, some considerations of how to implement assessment guidelines in a classroom will be addressed. The ways described are by no means exhaustive, but rather, they represent an attempt to draw more attention to looking at potential applications. Teachers hold the best position to make assessment decisions and most of the responsibilities for implementing assessment standards in the context of teaching.

📖 *Planning Instruction and Assessments*

While planning instruction, also be planning assessments that are a part of the instruction and thinking about culminating assessments for the instructional units. The assessments should reflect the kinds of learning opportunities students have had, and assessments should also provide opportunities to move toward higher cognitive levels, where students engage in assessments beyond recall and comprehension. The following strategies may be useful when planning:

- Identify big ideas and subconcepts that are pertinent to the school or district framework for science. Identify possible science-related issues that could address these concepts within the community. The local relevancy that this can provide for learning science also can be used to frame assessments that involve students in using critical thinking and problem-solving skills.

- Identify the assessment strategies and approaches that have potential for alignment with instruction. Are there performance tasks that can be used as assessments? How can instructional tasks be modified for use as performance tasks? Sound instructional tasks also tend to be sound performance tasks.

- Decide on several major assessment tasks that students will perform during the learning process, such as projects, group discussions, demonstrations, or presentations.

- Consider instructional and assessment possibilities that provide opportunities that address attitudes, creativity, and the nature of science.

- Select assessments that align with the kinds of information needed about student performance. Multiple-choice items are efficient for checking knowledge and comprehension, but higher-order thinking skills may call for other kinds of questions.

Assessment Reminders for Educators

Remember the following six points:

1. A given assessment is only one piece of information used to provide evidence for meeting teaching goals. Assessment should be

considered a tool in the service of the learner rather than as a teacher effectiveness measure.

2. The integration of assessment tasks into real-world problem-solving contexts should be a goal of assessment. Multiple assessments are recommended, and an emphasis should be placed on monitoring students' performances throughout the learning process and not just in the culminating assessment.

3. When looking at student outcome measures, questions that should be raised are those of when and where students have had the opportunity to experience and learn the information being assessed. A responsible teacher needs to be aware of and in control of as many factors as possible to reach a fair judgment when a given assessment is used.

4. Assessment standards are typically set by individuals other than students, but the teacher can at least invite students to participate in the establishment of some criteria for their assessments. Students should have an opportunity to contribute what they agree to learn and should learn how to assess their own learning.

5. Peer evaluation, group performance, and participation can be considered in determining a final grade.

6. In addition to normative comparisons, personal growth and ability levels should be taken into account in some ways when assessing students. For students to become engaged in learning, the opportunities to experience some level of success would seem to be necessary in classroom instruction and assessment.

Embedding Assessment in Teaching Practice

While planning instruction and assessments, it is helpful to have guidelines for moving ideas into practice. Some of the following strategies can help guide the planning process:

Communication of Expectations 📖

- Teacher expectations for assessment and how the curriculum will be implemented should be clearly conveyed and discussed with students. For example, teachers should discuss with students the instructional goals and evidence of learning, the kinds of data or works to be collected, how student performance will be assessed, and when student work is due.

- Student participation in the planning of what is to be learned and assessed is essential to promoting self-regulation of learning. Real-world, problem-solving activities, rather than textbook or memorization-oriented tasks, are probably necessary to have meaningful student involvement in learning.

Communication and Monitoring During the Learning Process

- Make certain the flow of curriculum is flexible enough to cope with the interaction in the classroom and retain the "how, what, and when" design initially indicated.

- Maintain a supportive learning atmosphere.

- Many assessment methods can be used to document student learning and modify the curriculum accordingly. For example, teacher observations of students could be used to monitor engagement.

- Students need time to internalize learning expectations, especially if they have not experienced more student-centered and open-ended learning environments.

Final Interpretation of Student Performance

- Validity and reliability of the assessments used to interpret and make decisions about student performance should always be scrutinized.

- As the teacher, be certain that claims made about student performance are based on both sound assessments and data interpretation.

- Teachers should use the information collected from student evaluations about what students perceive they are learning and doing. This feedback from students can inform the teacher in facilitating a student-centered classroom. Ideally, if students are learning in ways that are best for them, student performance should improve.

The World Wide Web as an Assessment Resource

The World Wide Web has a wealth of information on assessment. The user must be a discerning practitioner in the selection of assessments or ideas to be used in the classroom, but many excellent sites exist. The 10 regional educational laboratories are excellent beginning points for a variety of assessment information (see Table 2.3).

With the current emphasis placed on assessment, many states have developed web sites that include samples and exemplars for assess-

TABLE 2.3   Regional Educational Laboratories

Name	Web Site
Northeast and Islands Regional Educational Laboratory at Brown University	http://www.lab.brown.edu
Laboratory for Student Success	http://www.temple.edu/LSS
Appalachia Educational Laboratory	http://www.ael.org
Southeastern Regional Vision for Education	http://serve.org
Southwest Educational Development Laboratory	http://www.sedl.org
North Central Regional Educational Laboratory	http://www.ncrel.org
Midcentral Regional Educational Laboratory	http://www.mcrel.org
Northwest Regional Educational Laboratory	http://www.nwrel.org
WestEd	http://www.wested.org
Pacific Resources for Education and Learning	http://www.prel.hawaii.edu

ments. The NSES can be accessed via http://www.nsta.org. For evaluation and assessment information, the Educational Resources Information Center (ERIC) web site is http://ericae.net/.

3

Evaluating Teaching Practice

The possibilities of ways to evaluate teaching practice are many: action research, video tapes, journals, instruments to evaluate classroom practice, science as inquiry, and more.

Action Research

Teachers are in a central position to determine the direction and success of educational reform, and with educational reform linked to teaching practice and professional growth, this requires the commitment of teachers. Teachers must begin or continue to inquire about their own practice, and action research can be used to do so. Action research, as Hopkins (1993) notes, combines a substantive act with a research procedure. It is the engagement in an inquiry with a personal attempt to understand the situation and to improve or reform the situation.

Gay and Airaisian (2000) indicate that four beliefs underlie action research. Action research provides an opportunity for teachers to engage in professional growth, and teachers who wish to improve their practice need data to do so. In the process, teachers can use research findings to improve practice: Teachers can examine research findings and make applications in their own settings.

When teachers collect data to answer questions about effective teaching practices, they function as researchers engaged in inquiry. In education, many people categorize this as action research, which means teachers can systematically investigate ways by which they can improve their teaching and student learning. The *action* component is taken by the teacher in striving to analyze various aspects of teaching and learning. The term *research* indicates that a research plan has been structured and ultimately could be made available for public critique in the form of professional publications or presentations. Alternatively, the findings could be used to document personal professional growth for teacher evaluations and self-improvement plans. Engaging in classroom action research means that teachers would develop study plans that involved collecting data related to students' achievement, opportunities to learn, and teaching practices.

When teachers conduct action research, many research questions could be posed, and numerous strategies for data collection are possible. The teacher should work to refine and focus the research question to be addressed. If questions are too global, the action research process can become an insurmountable task. A highly focused question and good research design to answer the question is imperative for success. A recommendation for anyone initiating action research would be to keep the research manageable within the context of the daily classroom activities.

One method available to examine student learning in the classroom could be videotaping where the camera can become the eyes for the teacher who wishes to record student actions for later review and critique. Teachers may miss student performances because other classroom duties take priority. Although audio recordings could also be used in a similar way, some advantages of video use that have been identified by teachers include the opportunities to view and critique components in the following list:

- Student attentiveness

- Student self-paced work

- Student-student interactions

- Student performance using science processes (observation, classification, measuring, communication)

- Student creativity (project work, adaptation of new ideas or processes, generation of alternative hypotheses)

- Student group participation

- Teacher-student interaction

- Teacher involvement in student performances

- Teacher questioning strategies

- Teacher presentation and facilitation skills

- Teacher self-evaluation and reflection

- Identification of assessment opportunities

- Foci for discussion with students

The videotape, along with other classroom artifacts such as photographs of classroom activities, journals, learning logs, and audiotapes, can become part of a teaching portfolio. The development of a set of written criteria to frame the use of information from videotapes would be recommended. This can provide a focus for documentation of events that can serve as benchmarks to which future data can be compared. These criteria can be set out in the form of rubrics or checklists, and some instrumentation is available already in the literature or even on-line. A number of classroom learning environment questionnaires can also serve as tools to explore how students respond to teaching practices. For the teacher, this type of action research can become a longitudinal study of teaching practice.

Engaging in classroom research for self-improvement is likely not something to which students' parents and guardians will object, but the teacher may want to inform parents and guardians of this through a letter indicating intent and how the information will be used. Students are human subjects, and certain permission and approval may be required when conducting this kind of research, especially if done in conjunction with a university or college. Confidentiality of information should also always be considered.

Teacher journaling is another tool that can help teachers reflect on which or how content pedagogical strategies have been implemented in the classroom. Teacher journal writing encourages thinking and can even serve the important function of integrating course content, self-knowledge, and practical experiences with teaching and learning situations. Journal entries can also provide evidence for a longitudinal professional growth profile. To be useful as a reflective tool, journaling probably needs to become a habit so the events of reflection and levels of reflection can indicate patterns and changes. Journal entries could focus on (a) personal beliefs and knowledge about teaching and learning; (b) student responsiveness to the instructional strategies used in class; (c) classroom applications of information learned through literature, colleagues, and inservice workshops; and (d) personal reflections and feelings about the teaching and learning process.

A teacher journal related to constructivist teaching experiences framed in an STS approach is shown in Figure 3.1 as an example. The STS unit focused on nutrition and the heart. A self-review of the entries can also provide insight on how reflective a teacher has been, and questions to frame the analysis might include, Is what I have written only a description of activities that occurred, or is what I have written really reflective? In what ways are the entries reflective?

FIGURE 3.1. Reflective Teaching Journal: An Example From Practice

Entry 1: Our class is doing a nutrition-heart unit for our 20-day module. We discussed a food guide pyramid (how many servings are needed, etc.). They were given a copy of the pyramid to keep. We will be making a booklet of all our information throughout our unit. They seemed pretty excited about the topic. Hopefully it will stay that way!

Entry 2: I gave the class a chance to come up with any ideas about nutrition that they would like to do or learn about. I was kind of disappointed, because they didn't really come up with anything original or helpful. I have lots of ideas myself, so hopefully we'll do fine! They took a pretest on general nutrition-heart questions. The results weren't too bad, but I will be looking for lots of improvement and for them to become more knowledgeable.

Entry 3: We went into more detail with the five food groups (the required servings, etc.). They got into groups and chose their favorite pizza. Then, they came up with what food groups and servings were in it or on it. We all shared the results.

Entry 4: The students are going to record what they eat for 3 days, starting today. They filled out their own food pyramid to refer to, also. I'm finding I don't have as much time to get everything done like I'd like to in these class periods.

Entry 5: We are keeping and comparing the students' 3-day eating record to the pyramid. Nearly everyone found some room to improve on at least 1 of the days! We started discussing physical fitness.

Entry 6: We discussed different types of physical activities and their favorites and how it's good to do these for fitness. They got into groups and had a list of various activities. They marked boxes as to whether that activity falls under warm-up, strength, endurance, flexibility, or cool down.

Entry 7: We had a little more discussion about physical fitness. They also wrote on a paper what activities they did with their families, then, what they might add to their activities with their families, and what they could try.

Entry 8: We are including self-esteem with nutrition and fitness. I gave them a chance to be specific about their strengths and accomplishments. They filled out a poster on "What I Like About Me," "What I'm Good At," and "Things That Make Me Proud." We shared a few things with the class.

Entry 9: The students made a wellness pyramid to be "tossed" like dice. They would toss it and select a challenging question to ask or answer, to review for a quiz on Monday.

(continued)

FIGURE 3.1. Reflective Teaching Journal: An Example From Practice *(continued)*

Entry 10: We didn't have school yesterday because of weather, so we had our quiz today on the food guide pyramid, and so on. I'm anxious to grade it tonight to see how they did. I just finished grading the quiz; I was very pleased with the results. Everyone improved on his or her score from the beginning pretest. I've got lots more I want to cover. I hope I have enough days to do it in!

Entry 11: I videotaped myself today during class. It went pretty well. I think the kids were more excited about it than I was! They divided up into groups of five and made two school lunch menus to be chosen during 1 week of school next month. They had to follow the requirement of having all five food groups in their menu. They were enjoying it!

Entry 12: They continued working on their menus. We got together and chose which ones to use and also discussed it with our cook. From February 7-11, we'll get to have these original menus.

Entry 13: We began talking about the heart (its purpose, parts, etc.). We watched a video on how the heart works and discussed it afterwards. They had a fun sheet about the heart and what we talked about after that. Also a tape on "Interviewing a Heart."

Entry 14: I videotaped a little bit today on the heart. We had some discussion and used posters and did a sheet for their booklets. I found out I use my hands a lot and could give more waiting time for the questions I ask my students.

Entry 15: I videotaped myself for my final part on tape. We discussed more on the heart and did a demonstration, which they got to participate in. We showed how a valve in your heart works, by squeezing a demonstrator. We then used a siphon and "blood" and showed how your heart pumps blood, how it can pump faster when exercising, how it always goes one way, and so on. They had fun with it! I'm finding I have many more things to cover and not enough time! I know I will go over 20 days in this!!

Entry 16: We practiced taking our pulses—resting and exercising. Some were surprised at the difference. I'm working on getting a speaker to come in and talk to the class. That has been difficult for me, to get a hold of someone—I'm still trying, though. The kids didn't know of anyone, so it was up to me! We studied for a quiz on the heart tomorrow. They know their information well!

Entry 17: We took our heart quiz. We also decided to put smoking into the heart-nutrition unit. We took a pretest to see what they knew. We also did a "tar" demonstration to see how it collects in their lungs (poured dark syrup into a measuring cup—they said when; they predicted how much collected in a person's lungs in a year).

Entry 18: We watched a "smoking" video. We did a Mystery Bag Activity. They had to guess what was in the bag by asking two students clue questions. They answered yes or no until they guessed what was in the bag by gathering information and narrowing possibilities. Which is the same thing scientists and doctors have to do to figure out what causes heart disease, or to figure out how smoking affects your heart.

Entry 19: We took the smoking pretest again—they scored much, much better after all of our discussing and other activities. We related back to eating right and nutrition and talked about snacks (what's good for you, etc.). They divided up into groups and talked about healthy snacks.

FIGURE 3.1. *(continued)*

Entry 20: The groups are coming up with two original snacks of their own. We will be making these into a recipe booklet and also trying some of them—sampling! (Yum!) They were expected to mark off which food groups the ingredients came from, also. They enjoyed working on them.

Entry 21: We each had a partner today and were "Racing Raisins." (It dealt with carbon dioxide.) The raisins were placed in soda pop. After sinking, they watched how many times they got to the surface. (They could do what they wanted to help it get to the surface, if they wanted to.) It was a challenge.

Entry 22: We were all Science "Fizz Whiz" people! We were in pairs and coming up with how soda is different from all other liquids (has carbon dioxide). We then did an activity where they put different things in soda (sugar, salt, sand, cotton swab, etc.) and watched what happened. They wrote or recorded what happened after each one. We discussed it afterwards. The kids thought it was great!

Entry 23: We collected menus from various restaurants. We compared them, seeing if they had all the food groups somewhere, what kinds of foods were available, and so on. They got into groups and talked about this and then reported to the class on the different ones. They learned which nutritional restaurants to go to!

Entry 24: Groups again! They decided which food group to use and are building a person using foods from that group. They will bring the ingredients needed in a few days. Continued menu reporting, also.

Entry 25: Our speaker came today and talked about nutrition and the heart. She had had bypass surgery and knew lots about nutrition. The kids were interested and asked lots of questions. I wanted a dietitian but couldn't find one. Next month, a nurse will be coming, which should prove to be valuable also. She's planning on taking blood pressures, and so forth. We'll look forward to that.

Entry 26: The foods were brought for the person they'll make. They constructed them today and had fun. We had six different ones. They only used either the fruit or vegetable group. I thought we might have more of a variety. They turned out well. We will show them to the preschool tomorrow. We worked on the skits, also.

Entry 27: The preschool loved the "food persons." Gave them some good examples for snacks. We also worked on our skits. They will be presented in a couple of weeks. The menus they made up were eaten at lunch all week this week. The kids were famous!

Entry 28: We played some nutrition games using what we learned. We reviewed other things for a miniquiz.

Entry 29: We had a miniquiz. They did wonderfully. We also played a game as a review. They then "webbed" and put up everything they'd learned. It was wonderful compared to when we'd started. I'm wrapping up my journal—we have a few loose ends to cover, but will finish up eventually. It's been fun; lots of work involved, but it was worth it. I'm ready to start another unit of STS.

Assessment of Classroom Practice With Videotaping

The use of videotapes to inform personal teaching perspectives aligns well with the construct of the teacher as a reflective practitioner. Videotapes of class lessons can provide feedback to inform teaching practice from self-evaluation of the videotapes. Peer review of videotapes also can be a very powerful tool for constructive feedback, and these tapes can also provide an opportunity for others to see examples of exemplary teaching. The forms that follow (Forms 3.1, 3.2, and 3.3) can be modified to conform to classroom needs. For example, if the teacher wished to focus on the kinds of questions being asked, he or she could provide descriptors and codes of question categories and then just enter the code and tally in the matrix.

Sources of Instrumentation for Assessing Practice

Most school districts will have some types of instruments that are used to evaluate teaching performance, and if they have not already done so, this may be the context in which teachers begin. Using the district evaluation tools can help teachers provide better evidence of how criteria are met and interpreted. Furthermore, this can lead to selection of better instrumentation on which to be evaluated. Universities and colleges typically have protocols and instruments that they use in preservice programs, and these may be made available for use by contacting the education department.

The Expert Science Teacher Educational Evaluation Model, developed by Burry-Stock (1993), includes a number of instruments designed to evaluate various facets of constructivist practice. The Interstate New Teacher Assessment and Support Consortium (INTASC) standards, while targeting new teachers, also provide a framework for teaching practice that could be used to evaluate the classroom practice of veteran teachers. INTASC standards are on-line at various sites, and to access these, use one of the search engines and type in INTASC. Various instruments that can be used in the classroom are available on-line, and conducting an on-line search by topic or key words is another option.

Self-Assessment of Constructivist Practice

As Brooks and Brooks (1993) note, it sounds like a simple proposition that we construct our own understandings of the world, but moving constructivism into teaching practice may take time and teacher awareness of indicators of constructivist practice. Forms 3.4, 3.5, and 3.6 may be useful in raising awareness in the move to a more constructivist classroom.

Science-as-Inquiry Surveys

As a teacher, your responses to the Science as Inquiry Survey (Form 3.7) can provide evidence and documentation for your perceptions of your classroom practice. The student form of the survey (Form 3.8) provides the opportunity for your students to input their perceptions. Comparisons can be made, and the data can be used to profile the science learning opportunities from both student and teacher perspectives. The NSES and *Benchmarks for Science Literacy* were referents for the development of these survey forms.

FORM 3.1

Assessment of Classroom Interactions
From Videotaping

Teacher Name: _____

Videotape Context	Time Spent Dispensing Information	Time in Front of Classroom	Time With Individual Students or Groups	Use of Student Questions: Kinds and Levels of Questioning
Tape 1: Date, Context				
Tape 2: Date, Context				
Tape 3: Date, Context				
Tape 4: Date, Context				
Category Average, Summary Comments				

Comments/Reminders::

FORM 3.2
Finding Evidence of Effective Teaching Practices
in Classroom Videos

Date and Lesson Context:

1. Expectations for Student Learning: Reminders to Self:
   Yes   No   Were expectations for what students were to know
                   and be able to do clearly conveyed?
                 What evidence supports the observations?

2. Questioning Strategies:
   Yes   No   Were various levels of questioning used?
                 What levels of questions were asked?
   Yes   No   Did the questions elicit student thinking at various
                   cognitive levels?
   Yes   No   Was wait time practiced?
   Yes   No   Were students engaged in asking questions relevant
                   to the class activities?
   Yes   No   Did all students have the opportunity to respond to questions?
                 How did I respond to student answers?
                 What kind of feedback did I provide?

3. On-Task Behavior:
   Yes   No   Were the students on task?
                 What evidence supports these observations?
   Yes   No   Was the room arrangement conducive to on-task behavior?
                 What evidence supports these observations?
                 If students were not on task, what changes could be made
                   in this lesson in the future?

4. Use and Modeling of Instructional Strategies:
   Yes   No   Were appropriate instructional strategies used
                   during the lesson?
   Yes   No   Were appropriate instructional strategies modeled during
                   the lesson?
   Yes   No   Were multiple learning styles addressed during the lesson?
   Yes   No   Did students have the opportunity to discuss what was
                   going on in the class?
   Yes   No   Did students have the opportunity to write about what
                   was going on in the class?

5. Student Assessment:
   Yes   No   Was assessment embedded in the lesson?
                 What kinds of feedback were provided for students?
                 What kinds of assessments were used?

Enger, S., & Yager, R. *Assessing Student Understanding in Science.* © 2001 by Corwin Press, Inc.

FORM 3.3

Self-Review of Classroom Teaching Video

Date and Lesson Context:

For each of the sections, A, B, and C, record the number of times the following were observed during each of two selected 10 minute periods

A. Initiatory, informational
Talking:
(discussion, lecture, or directions)

(Early 10 min.)		(Late 10 min.)	
Yes/No	Open-Ended	Yes/No	Open-Ended

Questions: (initiatory)

Notes to self:

B. Responding

Teacher centered
(rejects or accepts student comments, confirms answer, repeats question or comment, clarifies or interprets, answers question directly)

Student centered
(asks students to clarify and/or elaborate)

Teacher-facilitated extensions (teacher discussion extending from student comment or question)

(Early 10 min.)		(Early 10 min.)	
Yes/No	Open-Ended	Yes/No	Open-Ended

Notes to self:

FORM 3.3 Continued

C. Wait time
(after the question)
teacher ↔ student

(Early 10 min.)		(Late 10 min.)	
Yes	No	Yes	No

Wait Time 2 (after the
first student response)
student ↔ student

Notes to self:

(Record the number of students on or off task at as many of the following intervals as possible during the observation.)

D. On or
off Task

On or off task

10 min.	15 min.	20 min.	25 min.	30 min.	35 min.	40 min.	45 min.

Improvement
targets:

SOURCE: Adapted from the SATIC coding form developed by Varrella, Kellerman, & Penick (1993).

FORM 3.4

Are You a **Constructivist** Teacher?

Teacher _____   Date _____

For statements 1 through 12, mark the letter that best indicates your perception of your teaching.

Use the following categories: a = Never, b = Seldom, c = Sometimes, d = Often

---

1. I encourage and accept student autonomy and initiative.   a   b   c   d

2. I use raw data and primary sources along with manipulative, interactive, and physical materials.   a   b   c   d

3. When framing tasks, I use cognitive terminology, such as "classify," "predict," and "create."   a   b   c   d

4. I allow student responses to drive lessons, shift instructional strategies, and alter content.   a   b   c   d

5. I inquire about students' understanding of concepts before sharing my understandings of those concepts.   a   b   c   d

6. I encourage students to engage in dialogue, both with other students and with me.   a   b   c   d

7. I encourage student inquiry by asking thoughtful, open-ended questions, and I encourage students to ask questions of each other.   a   b   c   d

8. I seek elaboration of students' initial responses.   a   b   c   d

9. I engage students in experiences that might engender contradictions to their initial hypotheses and then encourage discussion.   a   b   c   d

10. I allow wait time after posing questions.   a   b   c   d

11. I provide time for students to construct relationships and create metaphors.   a   b   c   d

12. I nurture students' natural curiosity through frequent use of the learning cycle model.   a   b   c   d

SOURCE: Developed from the work of Brooks and Brooks (1993).

Note: Complete this self-perception survey at the beginning of the year, at mid-year, and at the end of the year. This can provide a self-check and target areas that you may want to change. Providing examples or scenarios from practice to support perceptions could be a very strong evidential accompaniment to this self-perception survey.

FORM 3.5

Student Instructional Preferences and Self-Motivation Survey

Name: _____ Date: _____

Class: _____

This survey asks you to give your preferences about how you like to learn science and what motivates you in science. Circle the letter that best matches your viewpoint.

Use the following categories: a = Strongly Agree, b = Agree, c = Disagree, d = Strongly Disagree

1. I prefer that my teacher tell me about science rather than to read a science book.   a   b   c   d

2. I like it when I have to explain the results of my own experiment.   a   b   c   d

3. Studying alone, I learn more than by studying in a small group.   a   b   c   d

4. In science classes, I would rather listen to the teacher than do other activities.   a   b   c   d

5. I like to do experiments, which help me to understand the science I have learned.   a   b   c   d

6. I like the teacher to explain rather than have to learn from books.   a   b   c   d

7. Taking tests helps me know if I have understood what I have learned in class.   a   b   c   d

8. I find it difficult to listen to the teacher for a long period of time.   a   b   c   d

9. I understand science concepts better if I have to explain them in my own words.   a   b   c   d

10. I like working in small groups in science.   a   b   c   d

11. I like the science teacher to decide how we learn science.   a   b   c   d

12. I learn more from doing experiments than by listening to the teacher's explanations.   a   b   c   d

13. I feel confused when I read several books about the same science idea.   a   b   c   d

14. I like to have my science teacher correct my homework.   a   b   c   d

15. I prefer to listen to the teacher rather than learn from doing experiments.   a   b   c   d

16. I like to find out something without the teacher telling me how to do it.   a   b   c   d

---

Enger, S., & Yager, R. *Assessing Student Understanding in Science.* © 2001 by Corwin Press, Inc.

(continued)

FORM 3.5 Continued

17. One of the best ways for me to understand science is to discuss it in class.   a   b   c   d

18. I would learn more if I could choose which science topics I studied.   a   b   c   d

19. I find it difficult to do science experiments without instructions from the teacher.   a   b   c   d

20. When I am interested in a scientific idea, I like to read more about it.   a   b   c   d

21. I would rather be tested by the teacher than anyone else.   a   b   c   d

22. I find it difficult to understand science without the teacher's explanations.   a   b   c   d

23. I would rather find out about a scientific idea on my own than have it explained by the teacher.   a   b   c   d

24. When working in small groups, my classmates share with me what they know.   a   b   c   d

25. I like the teacher to tell me what I have to do when doing an experiment.   a   b   c   d

26. The best science classes are those when we do experiments.   a   b   c   d

27. Taking notes is more useful for learning than reading textbooks.   a   b   c   d

28. My classmates know better than the teacher does whether or not I understand science.   a   b   c   d

29. By taking notes, I make sure that I study what the teacher wants me to learn.   a   b   c   d

30. Solving problems is one of the best ways for me to understand science.   a   b   c   d

31. I express my ideas more easily when I am working in a small group.   a   b   c   d

32. The teacher's answers to the questions asked in class by my classmates help me understand science.   a   b   c   d

33. I enjoy doing experiments.   a   b   c   d

34. I would rather use computers to learn science than to listen to the teacher.   a   b   c   d

35. Taking a test is not the only way of finding out if I have understood science.   a   b   c   d

36. I like to get good grades, even if I have to work hard to do so.   a   b   c   d

FORM 3.5 Continued

37. In science classes, if I do not understand something, I look it up in a book.   a   b   c   d

38. I get worried if I cannot solve a problem in science.   a   b   c   d

39. I would rather have friends than to be the best in the class.   a   b   c   d

40. I do not like other classmates to know if I get a poor grade.   a   b   c   d

41. I like learning about the latest discoveries and inventions in science.   a   b   c   d

42. I do not mind working hard in science class as long as I learn something.   a   b   c   d

43. I care what my classmates think of me.   a   b   c   d

44. I like to compete with others for the best grades.   a   b   c   d

45. In the science lab, I like to mix different chemicals to find out what happens.   a   b   c   d

46. I am ashamed when I get a low grade on a test.   a   b   c   d

47. When doing experiments, I prefer to work with my friends.   a   b   c   d

48. I like the teacher to tell the rest of the class when I get good grades.   a   b   c   d

49. I like to find out about new ideas in science.   a   b   c   d

50. I like the teacher to praise my efforts in science.   a   b   c   d

51. Having good friends is one of the most important things at school.   a   b   c   d

52. I try to lead in class discussions.   a   b   c   d

53. I am interested in many scientific ideas that are not taught at school.   a   b   c   d

54. I try to pay attention to what the teacher says so that I will not miss anything important.   a   b   c   d

55. I do not mind it when classmates copy my problems or work.   a   b   c   d

56. I like to get the best grade on tests.   a   b   c   d

57. I like to find out more information than what the teacher tells me in class.   a   b   c   d

58. I like it when the teacher gives detailed explanations.   a   b   c   d

59. I do not mind lending my books and notes to classmates.   a   b   c   d

60. I like to be one of the first to finish my class work.   a   b   c   d

---

Enger, S., & Yager, R. *Assessing Student Understanding in Science.* © 2001 by Corwin Press, Inc.

(continued)

FORM 3.5 Continued

61. I want to know more about many new science topics.	a	b	c	d
62. I like the teacher to check my homework every day.	a	b	c	d
63. I like my classmates to help me in class.	a	b	c	d
64. I am more interested in the grade I get than in the mistakes I have made.	a	b	c	d
65. I am interested in finding out the answers when solving scientific problems.	a	b	c	d
66. I try hard to please the teacher with my work.	a	b	c	d
67. I like working with friends when working in small groups.	a	b	c	d
68. In class discussions, I like to be able to present the best ideas.	a	b	c	d
69. I enjoy reading books about science.	a	b	c	d
70. I like to do my best when doing my science homework.	a	b	c	d
71. I like working in small groups.	a	b	c	d
72. I like to show others my answer when they do not know how to do the work.	a	b	c	d
73. I like to learn about new ideas in science.	a	b	c	d
74. I like homework because I learn more.	a	b	c	d
75. When working in small groups, I do not care with whom I work.	a	b	c	d

SOURCE: Adapted from Giddings (1993).

FORM 3.6

What Happens in My Science Classroom? Student Form

Questionnaire purpose:

This questionnaire asks you to provide your opinions about your science classroom.
This is not a test, and your answers will not affect your grade. There are no right or wrong answers.
Your answers will help your teacher improve your science classes.

For each sentence, circle the letter in the column at the right which best describes you or your views.

Learning About the World	Almost Always	Often	Some-times	Almost Never
In this class,				
1. I learn about the world outside of school.	a	b	c	d
2. My new learning starts with problems about the world outside of the school.	a	b	c	d
3. I learn how science can be a part of my out-of-school life.	a	b	c	d
In this class,				
4. I get a better understanding of the world outside of school.	a	b	c	d
5. I learn interesting things about the world outside of school.	a	b	c	d
6. What I learn has *nothing* to do with my out-of-school life.	a	b	c	d

Learning About Science	Almost Always	Often	Some-times	Almost Never
In this class,				
7. I learn that science *cannot* provide perfect answers to problems.	a	b	c	d
8. I learn that science has changed over time.	a	b	c	d
9. I learn that science is influenced by people's values and opinions.	a	b	c	d
In this class,				
10. I learn about the different sciences used by people in other cultures.	a	b	c	d
11. I learn that modern science is different from the science of long ago.	a	b	c	d
12. I learn that science is *inventing* theories.	a	b	c	d

(continued)

FORM 3.6 Continued

Learning to Express Myself	Almost Always	Often	Some-times	Almost Never
In this class,				
13. It's OK for me to ask the teacher "why do I have to learn this?"	a	b	c	d
14. It's OK for me to question the way I'm being taught.	a	b	c	d
15. It's OK for me to complain about activities that are confusing.	a	b	c	d
In this class,				
16. It's OK for me to complain about anything that prevents me from learning.	a	b	c	d
17. It's OK for me to express my opinion.	a	b	c	d
18. It's OK for me to speak up for my rights.	a	b	c	d

Learning to Learn	Almost Always	Often	Some-times	Almost Never
In this class,				
19. I help the teacher plan what I'm going to learn.	a	b	c	d
20. I help the teacher decide how well I am learning.	a	b	c	d
21. I help the teacher decide which activities are best for me.	a	b	c	d
In this class,				
22. I help the teacher decide how much time I spend on activities.	a	b	c	d
23. I help the teacher decide which activities I do.	a	b	c	d
24. I help the teacher assess my learning.	a	b	c	d

FORM 3.6 Continued

Learning to Communicate	Almost Always	Often	Some-times	Almost Never
In this class,				
25. I get the chance to talk to other students.	a	b	c	d
26. I talk with other students about how to solve problems.	a	b	c	d
27. I explain my ideas to other students.	a	b	c	d
In this class,				
28. I ask other students to explain their ideas.	a	b	c	d
29. Other students ask me to explain my ideas.	a	b	c	d
30. Other students explain their ideas to me.	a	b	c	d

SOURCE: Adapted from the work of Taylor, Fraser, and White (1994).

Enger, S., & Yager, R. *Assessing Student Understanding in Science.* © 2001 by Corwin Press, Inc.

FORM 3.7

Science As Inquiry: Teacher Perceptions of Science Class

The questions on this survey relate to elements of teaching practice in your science classroom.

SECTION 1: SCIENCE CLASSROOM

Use the following rating scale for questions 1-34: 5 = Very Often, 4 = Often, 3 = Sometimes, 2 = Seldom, 1 = Never

In your science class, how often do you have your students do the following?	Very Often	Often	Sometimes	Seldom	Never
1. Work in groups or teams	5	4	3	2	1
2. Work in groups or teams when they do science activities	5	4	3	2	1
3. Work individually	5	4	3	2	1
4. Do activities and experiments in science class	5	4	3	2	1
5. Work alone to do science activities and experiments	5	4	3	2	1
6. Design their own activities and experiments	5	4	3	2	1
7. Try activities or experiments that they design themselves	5	4	3	2	1
8. Test a hypothesis or question in their activities or experiments	5	4	3	2	1
9. Control variables when they do lab activities or experiments	5	4	3	2	1
10. Ask questions and then investigate their own questions	5	4	3	2	1
11. Make predictions about what will happen before they do activities or experiments	5	4	3	2	1
12. Set up a data table when they do activities or experiments	5	4	3	2	1
13. Make observations when they do activities or experiments	5	4	3	2	1
14. Write down their observations from an experiment	5	4	3	2	1
15. Write about the experiments that they do in a notebook, log, or journal	5	4	3	2	1
16. Write down their own information from a science experiment	5	4	3	2	1
17. Graph numbers from their experiments	5	4	3	2	1
18. Discuss the results from their experiments	5	4	3	2	1
19. Set up their own experiments or activities	5	4	3	2	1
20. Try experiments more than one time to check their results	5	4	3	2	1
21. Read about the research work that scientists do	5	4	3	2	1
22. Discuss the research work that scientists do	5	4	3	2	1
23. Discuss science articles from newspapers or magazines	5	4	3	2	1

Enger, S., & Yager, R. *Assessing Student Understanding in Science.* © 2001 by Corwin Press, Inc.

	Very Often	Often	Sometimes	Seldom	Never
24. Go to the school library or media center to find science information	5	4	3	2	1
25. Watch and then discuss science videos	5	4	3	2	1
26. Have visitors come to class to talk about science	5	4	3	2	1
27. Go on field trips on campus (school grounds) that relate to what they do in science class	5	4	3	2	1
28. Go on field trips off campus that relate to what they do in science class	5	4	3	2	1
29. I do experiments for the class.	5	4	3	2	1
30. If activities or experiments do not appear to work as predicted, we discuss reasons why.	5	4	3	2	1

Who decides what science lessons and activities are done in science class?

	Very Often	Often	Sometimes	Seldom	Never
31. I decide what the science lessons are about.	5	4	3	2	1
32. The students in the class decide what the science lessons are about.	5	4	3	2	1
33. I decide what science activities and experiments we do.	5	4	3	2	1
34. The students in the class decide what science activities and experiments we do.	5	4	3	2	1

SECTION 2: SCIENCE ASSIGNMENTS

Rating scale for Questions 35-44:   5 = Very Often (3 to 5 times a week),   4 = Often (1 or 2 times a week),
3 = Sometimes (1 or 2 times a month),   2 = Seldom (1 or 2 times a year),   1 = Never (not done in science class)

In your science class, how often do you have your students do the following?

	Very Often	Often	Sometimes	Seldom	Never
35. How often do you give science assignments?	5	4	3	2	1
36. Do students work in groups to complete some assignments?	5	4	3	2	1

What kinds of assignments do you use in your science class?

	Very Often	Often	Sometimes	Seldom	Never
37. Students answer questions at the end of a section or chapter in the science textbook.	5	4	3	2	1
38. Students write definitions of science words.	5	4	3	2	1
39. Students do science worksheets.	5	4	3	2	1
40. Students do concept maps, mind maps, or webs.	5	4	3	2	1
41. Students do assignments that require presentations.	5	4	3	2	1

(continued)

Enger, S., & Yager, R. *Assessing Student Understanding in Science.* © 2001 by Corwin Press, Inc.

FORM 3.7 Continued

What kinds of assignments do you use in your science class? (continued)	Very Often	Often	Sometimes	Seldom	Never
42. Students do assignments that require the use of a variety of resources.	5	4	3	2	1
43. Students do assignments that require projects.	5	4	3	2	1
44. Students do assignments that require them to involve people from the community.	5	4	3	2	1

SECTION 3: TESTS AND ASSESSMENTS

Rating scale for Questions 45-58: 4 = Frequently (for most tests or assessments), 3 = Sometimes (for some tests or assessments), 2 = Seldom (for very few tests or assessments), 1 = Never (not used in science class)

How often do you have the following on tests or for assessments?	Frequently	Sometimes	Seldom	Never
45. True-false questions	4	3	2	1
46. Multiple-choice questions	4	3	2	1
47. Matching questions	4	3	2	1
48. Fill-in-the-blank questions	4	3	2	1
49. Short-answer questions	4	3	2	1
50. Essay questions	4	3	2	1
51. Short-term products or projects (that take about 1 week to do)	4	3	2	1
52. Longer-term products or projects (that take more than 1 week to do)	4	3	2	1
53. Write reports about science	4	3	2	1
54. Maintain a science portfolio of their work	4	3	2	1
55. Make presentations about their work	4	3	2	1
56. Do concept maps, mind maps, or webs	4	3	2	1

Do you involve students in the following?

	Frequently	Sometimes	Seldom	Never
57. Use student input for design of a scoring guide (rubric) for their work	4	3	2	1
58. Use student input for decisions on how some science work is graded	4	3	2	1

Enger, S., & Yager, R. *Assessing Student Understanding in Science.* © 2001 by Corwin Press, Inc.

SECTION 4: EQUIPMENT AND MATERIALS USE

Rating scale for Questions 59-69: 5 = Very Often, 4 = Often, 3 = Sometimes, 2 = Seldom, 1 = Never

Do students use any of the following equipment or materials in science class?	Very Often	Often	Sometimes	Seldom	Never
59. Balances or scales	5	4	3	2	1
60. Thermometers	5	4	3	2	1
61. Microscopes	5	4	3	2	1
62. Magnifying lenses	5	4	3	2	1
63. Meters sticks or rulers	5	4	3	2	1
64. Timers or stopwatch	5	4	3	2	1
65. Computers for word processing	5	4	3	2	1
66. Computers with probes or science software	5	4	3	2	1
67. Live animals or plants	5	4	3	2	1
68. Preserved animals or plants	5	4	3	2	1
69. Graduated cylinders or containers to measure liquids	5	4	3	2	1

SOURCE: Enger, S. (1997).

Enger, S., & Yager, R. *Assessing Student Understanding in Science.* © 2001 by Corwin Press, Inc.

Science as Inquiry: Student Perceptions of Science Class

The questions on this survey relate to things that you may do in your science class when you are learning about science.

SECTION 1: SCIENCE CLASSROOM

Use the following rating scale for Questions 1–34:   5 = Very Often,   4 = Often,   3 = Sometimes,   2 = Seldom,   1 = Never

In your science class, how often do you do the following?	Very Often	Often	Sometimes	Seldom	Never
1. Work in groups or teams	5	4	3	2	1
2. Work in groups or teams when you do science activities	5	4	3	2	1
3. Work by yourself	5	4	3	2	1
4. Do activities and experiments in science class	5	4	3	2	1
5. Work by yourself when you do science activities and experiments	5	4	3	2	1
6. Design your own activities and experiments	5	4	3	2	1
7. Try activities or experiments that you design yourselves	5	4	3	2	1
8. Test a hypothesis or question in your activities or experiments	5	4	3	2	1
9. Control variables when you do lab activities or experiments	5	4	3	2	1
10. Ask questions and then investigate your own questions	5	4	3	2	1
11. Make predictions about what will happen before you do activities or experiments	5	4	3	2	1
12. Set up a data table when you do activities or experiments	5	4	3	2	1
13. Make observations when you do activities or experiments	5	4	3	2	1
14. Write down your observations from an experiment	5	4	3	2	1
15. Write about the experiments that you do in a notebook, log, or journal	5	4	3	2	1
16. Write down your own information from a science experiment	5	4	3	2	1
17. Graph numbers from your experiments	5	4	3	2	1
18. Discuss the results from your experiments	5	4	3	2	1
19. Set up your own experiments or activities	5	4	3	2	1
20. Try experiments more than one time to check your results	5	4	3	2	1
21. Read about the research work that scientists do	5	4	3	2	1

	Very Often	Often	Sometimes	Seldom	Never
22. Discuss the research work that scientists do	5	4	3	2	1
23. Discuss science articles from newspapers or magazines	5	4	3	2	1
24. Go to the school library or media center to find science information	5	4	3	2	1
25. Watch and then discuss science videos	5	4	3	2	1
26. Have visitors come to class to talk about science	5	4	3	2	1
27. Go on field trips on campus (school grounds) that relate to what you do in science class	5	4	3	2	1
28. Go on field trips off campus that relate to what you do in science class	5	4	3	2	1
29. The teacher does experiments for the class	5	4	3	2	1
30. If activities or experiments do not appear to work as predicted, we discuss reasons why.	5	4	3	2	1

Who decides what science lessons and activities are done in science class?

	Very Often	Often	Sometimes	Seldom	Never
31. The teacher decides what the science lessons are about	5	4	3	2	1
32. The students in the class decide what the science lessons are about	5	4	3	2	1
33. The teacher decides what science activities and experiments we do	5	4	3	2	1
34. The students in the class decide what science activities and experiments we do	5	4	3	2	1

SECTION 2: SCIENCE ASSIGNMENTS

Rating scale for Questions 35-44:   5 = Very often (3 to 5 times a week),   4 = Often (1 or 2 times a week),
3 = Sometimes (1 or 2 times a month),   2 = Seldom (1 or 2 times a year),   1 = Never (not done in science class)

In your science class, how often do you do the following?	Very Often	Often	Sometimes	Seldom	Never
35. How often do you have science assignments?	5	4	3	2	1
36. Do you work in groups to complete some assignments?	5	4	3	2	1

(continued)

FORM 3.8 Continued

What kinds of assignments do you have in your science class?	Very Often	Often	Sometimes	Seldom	Never
37. We answer questions at the end of a section or chapter in the science textbook	5	4	3	2	1
38. We write definitions of science words	5	4	3	2	1
39. We do science worksheets	5	4	3	2	1
40. We do concept maps, mind maps, or webs	5	4	3	2	1
41. We do assignments that require us to make presentations	5	4	3	2	1
42. We do assignments that require us to use a variety of resources	5	4	3	2	1
43. We do assignments that require us to complete projects	5	4	3	2	1
44. We do assignments that require us to involve people from the community	5	4	3	2	1

SECTION 3: TESTS AND ASSESSMENTS

Rating scale for Questions 45-58: 4 = Frequently (for most tests or assessments),   3 = Sometimes (for some tests or assessments),
2 = Seldom (for very few tests or assessments), 1 = Never (not used in science class)

How often do you have the following on tests or for assessments?	Frequently	Sometimes	Seldom	Never
45. True-false questions	4	3	2	1
46. Multiple-choice questions	4	3	2	1
47. Matching questions	4	3	2	1
48. Fill-in-the blank questions	4	3	2	1
49. Short-answer questions	4	3	2	1
50. Essay questions	4	3	2	1
51. Short-term products or projects (that take about 1 week to do)	4	3	2	1
52. Longer-term products or projects (that take more than 1 week to do)	4	3	2	1
53. Write reports about science	4	3	2	1
54. Keep a science portfolio of your work	4	3	2	1
55. Make presentations about your work	4	3	2	1

Enger, S., & Yager, R. *Assessing Student Understanding in Science.* © 2001 by Corwin Press, Inc.

Do you do the following?	Frequently	Sometimes	Seldom	Never
56. Do concept maps, mind maps, or webs	4	3	2	1
57. Help the teacher design a scoring guide (rubric) for your work	4	3	2	1
58. Decide on how some science work is graded	4	3	2	1

SECTION 4: EQUIPMENT AND MATERIALS USE

Rating scale for Questions 59-69: 5 = Very Often, 4 = Often, 3 = Sometimes, 2 = Seldom, 1 = Never

Do you use any of the following equipment or materials in science class?	Very Often	Often	Sometimes	Seldom	Never
59. Balances or scales	5	4	3	2	1
60. Thermometers	5	4	3	2	1
61. Microscopes	5	4	3	2	1
62. Magnifying lenses	5	4	3	2	1
63. Meters sticks or rulers	5	4	3	2	1
64. Timers or stopwatch	5	4	3	2	1
65. Computers for word processing	5	4	3	2	1
66. Computers with probes or science software	5	4	3	2	1
67. Live animals or plants	5	4	3	2	1
68. Preserved animals or plants	5	4	3	2	1
69. Graduated cylinders or containers to measure liquids	5	4	3	2	1

SOURCE: Enger, S. (1997).

Enger, S., & Yager, R. *Assessing Student Understanding in Science.* © 2001 by Corwin Press, Inc.

4

Rubrics and Scoring Guides

Assessment should include the use of criteria designed to communicate goals and desired learner outcomes to the student. *Rubric,* from the Latin for "red," refers to an established form or method, and rubrics or scoring guides can provide forms and methods for communicating criteria for activities, events, conceptual development, or goals for learners in the class. These frameworks provide a structure on which student work can be developed and assessed. Rubrics and scoring guides can be designed for a wide variety of situations, such as presentations, investigative projects, or performance tasks.

Criteria for student outcomes have been viewed many times as end-of-the-year goals are met, usually one section, unit, or chapter at a time. Criteria should allow teachers to develop end-of-unit assessments that indicate how each student is progressing toward the goal of understanding district and other desired learning outcomes. Current understanding and reforms in science education depict learning as an active process in which the students learn individually, collectively, competitively, and collaboratively. Each of the aforementioned characteristics is present in the successful classroom; yet they are often not assessed because of district requirements for standardized testing, time constraints, or personal beliefs about teaching and testing.

To assess the objectives represented by the type of active learning desired for students, student understanding should be assessed prior to instruction, during the instructional phase, and following the instructional sequence. Assessment should be viewed as an ongoing component of the classroom. Although many would argue that there is little time to do all of this assessment, a counterargument would be that each student is an individual, and no two students could possibly know exactly the same material or have the same mental constructs for any subject. For students to develop understanding, time must be focused on both formative and summative assessments, and the information collected should be used to inform instruction. The results of these assessments should then be used to develop the criteria each student will address prior to completing the instructional sequence. This is where rubrics or scoring guides can be used to focus thinking and learning for both the teacher and the student.

This chapter includes examples of rubrics that have been developed and used by teachers, and the teachers and their students have created these for their specific needs. The sampling of rubrics includes some that are related to biology, to concept mapping, and to physical science topics. The opportunities are limitless for adapting these to many learning situations. A selection of examples of not very good rubrics is included at the end of the chapter.

Rubric Construction: A Perspective From a Teacher

A scoring rubric or grid is an essential part of an assessment model. The model speaks to being up front with the students in terms of what is expected and at what level or standard the work should be completed. The rubric is based on expected outcomes, and the instructor asks him- or herself what he or she expects the student to be able to do after having completed the work assigned. These expectations become the outcomes that guide student products or performances. The expected outcomes become strands on the scoring rubric. Each of the strands indicates to the student the standards for exemplary work, acceptable work, and work that must be revised. The teacher might ask, "What is one outcome I might expect from a student writing a paper on a scientific investigation?" Perhaps one of the outcomes would be that the student made few scientific errors. As a result, the strand for that outcome might appear as follows:

3	*2*	*1*
No major errors in scientific investigation; high level of accurate information	Some scientific errors that distract the reader from the major significance of the information; some information inaccurate or omitted	Enough scientific errors to render the essay ineffective for useful information; important information highly inaccurate or omitted

The inclusion of one desired outcome in the descriptors may initially make the rubric easier for the instructor and students to use. The rubric intends to communicate to the student that accuracy of the investigation is of prime importance. The teacher would continue to build the rubric based on the desired outcomes for this particular assignment. The rubric could include other strands, such as insights the student was able to form based on completed research, quality of the research materials used, and ability to support a conclusion based on clear evidence. Each idea or outcome would create one strand of the rubric. The descriptions of the standards should be parallel across the score points. A strand on resource use for this rubric follows:

3	2	1
Indication of multiple use of resources; good use of research to support thesis	Research goes beyond basic text; indication of use of alternative viewpoints; some additional information would enhance support of writer's thesis	Little or no evidence of research beyond the basic textbook

It is important to include only one desired outcome in the descriptors for the standards in each strand. Otherwise, it becomes more difficult to decide how to award points if the student achieves one part of a multiple strand but not all parts. If the strand addresses only one idea, then the determination about achieving the standard is clearer. Setting out the rubric or scoring guide in advance provides a communication guide for both the instructor and the students.

The instructor should be certain that all outcomes to be assessed are addressed in the rubric, because it would be unfair to assess students on criteria not communicated by the rubric. If one reason for initially giving students the rubric is to demystify the way in which they will be assessed, it would be unfair to change the rubric without informing them. If a rubric has been prepared that does not address an outcome that is later thought to be desirable, then the rubric should be rewritten if the same assignment is to be given later. Rubric or scoring guide development tends to become an iterative process as descriptors are refined with repeated use of the rubric.

Rubrics may appear to be subjective, because words like "no major errors" rather than "fewer than 10 errors" are used. If a rubric has standards that indicate exemplary work as giving six causes of the Civil War, the student is expected to equate quantity with quality. He or she may give six causes, but are they meaningful causes? Are there more than six important causes? Is the student more interested in finding six causes than concentrating on their importance? It may be better to allow the student to decide for himself or herself how many important causes can be successfully supported as relevant. This allows the student to place more emphasis on quality of response rather than quantity.

Rubrics further serve the purpose of making the evaluation of papers more focused. Students can be clearer about teacher requirements and expectations, which in turn can lead them to write better papers. Teachers themselves also can be clearer about expectations. Therefore, reading the paper with the clear intent of evaluating already established criteria is likely to result in a fairer assessment of student work. No longer does the teacher read papers to discover the standards, but the standards for performance have been established at the outset. Students can also submit a self-assessment or their work based on the rubric.

In summary, a rubric can serve students in the following ways:

- Demystifies teacher expectations for the students

- Sets criteria and standards for performance in advance

- Allows the student to explain what he or she knows about a subject, rather than determine the right answer as identified by the teacher

- Gives the student a clearer picture of how to organize a paper or project

- Makes the evaluation of student materials more focused and objective

- Provides a self-assessment opportunity for the students

Rubric or Scoring Guide Design 📖

Before even beginning the design of a rubric or scoring guide, consider the assessment design. *Purpose* is the key that should guide the development of any assessment. What is the purpose of the assessment? Does assessment align with the learning experiences? What are the intended outcomes that can be expected from students as a result of their learning experiences? What are the measurable student outcomes that can realistically be expected? The use of a rubric or scoring guide can be a mechanism to communicate assessment intentions and to provide feedback to the students about their performances. This assessment information is often combined in some way with other information for grading and reporting purposes.

Rubric design and development can be accomplished in multiple ways. A general framework may be set up, and then this general framework can be customized for specific assessments. The reverse strategy can also be used, with a specific rubric designed for each assessment, which could then be generalized. The general template approach may work well, in that the overarching assessment framework is in place. The works of Nitko (1996) and Marzano, Pickering, and McTighe (1993) are two useful references for the design process, and with the current emphasis on assessment, many other excellent sources are available in print or on the web. Numerous states have and are developing assessment frameworks, and both assessment samples and scoring guides are on-line at state department of education web sites.

A general rubric grid could be organized like the example that shortly follows. Three categories may suffice to assess the performance; 5, 3, and 1 are often used to head the columns. A set of exemplars for each category can be collected over time and used to communicate to students what attributes exemplify each performance level. It is strongly recommended that if student work falls in the "not acceptable" category that constructive feedback be provided and where feasible, the student be provided an opportunity to resubmit the work. Equally important is the feedback to those whose work meets the "best" level so that the attributes can be met in future work.

Sample Grid for a Rubric

5 (Best)	3 (Good)	1 (Fair—Improvement Needed)
This block usually addresses what the teacher and students consider to be the best possible work that could be done by a student or group. All criteria are met.	Work that meets this level of success is good but does not meet all of the criteria set by the teacher and students. Some errors or flaws are present in the work.	Criteria in this block indicate that the work completed clearly shows that it could be improved on. This area addresses the kind of work, attitude, or achievement that is considered to be unacceptable by the teacher and students.

It is important for students to have some level of input in setting the criteria, because this act empowers students in their assessment. One strategy to gain students' input is to provide them with a draft rubric and have them provide or help write descriptors and explanations for performance levels. After this process students should have a better understanding of how to meet the expected outcomes.

Rubric for Interdisciplinary or Specified Domain Assessment[1]

The development of criteria and the resulting rubric are presented here. This style of rubric was designed for the assessment of student work from a biology class that used STS ideas.

This rubric includes elements that address expectations for the generation of a model, a presentation, or both. In rubric design, consider the number of categories to be established because as more categories are established, more score points are needed, and this sometimes makes it more difficult to establish descriptors of performance. Although the following rubric template is set out to have five categories, it could be collapsed to three with score points of 5, 3, and 1, which may make it easier to assess performance. When a product or project is complex, trying to isolate individual pieces can become very difficult.

Science Domain	*5*	*4*	*3*	*2*	*1*
Scientific thought (creativity, concepts, world view)	Original and creative design. Strong evidence of understanding the scientific principles of _____ *(insert topic or concepts here)*	Attempt at an original design; expresses some creativity, and provides some evidence of understanding the scientific principles	Model built from kit or plan; less creativity noted than with a more original design; some evidence of understanding scientific principles	Model built from kit or plan; creative elements not apparent; limited understanding of scientific principles	Model built from kit or plan; lack of creativity; appears to not understand scientific principles
Presentation (language arts, speech, conceptual understanding)	Clear and concise; effective use of scientific terms; complete understanding of topic for this age group; able to extrapolate	Well organized; good use of science terms; good understanding of topic; a few errors present	Presentation acceptable; adequate use of science terms; acceptable understanding of topic; more errors present	Presentation lacks clarity and organization; little inclusion of scientific terms and vocabulary; limited understanding of topic; numerous errors	No presentation developed; no intent to present
Exhibit (attitude, world view, use of technology, art)	Model completed to specifications; excellent workmanship; materials used in exemplary fashion Information is self-explanatory	Model completed to specifications; good workmanship; effective use of materials; information clear and logical	Model completed; adequate workmanship; materials used appropriately; acceptable information	Model partially completed; some problems with workmanship; materials could be used more effectively; some information lacking	Model not completed; poor workmanship; poor use of materials

Biology Assessment Task[2]

📖 *Performance Task:*

The U.S. Surgeon General has selected you and your colleagues to serve on a committee responsible for raising public awareness of an environmental issue that has a potential impact on human health. The U.S. Surgeon General would like you to identify an issue and develop a position statement and suggestions for addressing the issue. To assist you with developing your position statement and recommendations, the following guidelines should help focus your task:

- Identify an environmental issue that has the potential to have an impact on human life.

- Prepare a position paper that addresses this issue and suggest ways to address this issue.

- You should research the issue your group has chosen.

- You should describe how this issue affects the human body systems.

- You should address potential problems if this issue is not addressed or resolved.

- You should develop a plan to educate others about your findings.

Criteria	10 (Set your own exemplary standards)	8	6	4
Issue identified and group position stated		Significance of issue clearly presented with detail	Significance of issue clearly presented but more detail needed	Issue stated but little said about the significance
Linkage of environmental effects on body systems		Two effects discussed per system; three additional effects identified and discussed	Two effects discussed per system; two additional effects identified and discussed	Two effects discussed per system; two additional effects identified and discussed
Evidence and support		Extensive support data included	Adequate support data included	Minimal support data included

Criteria	10 (Set your own exemplary standards)	8	6	4
Body Systems				
Reproductive				
Respiratory				
Circulatory				
Nervous				
Other				
Plan to educate others		Complete education component for identified systems	Adequate education component for identified systems	Minimal education component for identified systems
		Creativity used to enhance means of education	Some creative endeavor attempted to enhance means of education	Minimum creative effort
Potential impact on life		Two examples of impact potential summarized per system; three additional examples included	Two examples of impact potential summarized per system; two additional examples included	Two examples of impact potential summarized per system; one additional example included
References (depth of research), reference page		Five different types of references; reference page included	Four different types of references; reference page included	Three different types of references; reference page included

The rubrics that follow are samples of teacher-developed rubrics, and these are likely to need modification to address the uniqueness of each classroom. The score points and weighting for each performance level can be added at the discretion of the teacher and students.

The Application Domain: A Proposal for Social Action[3]

Distinguished Performance	Proficient Performance	Intermediate Performance
Proposal has realistic and practical applications	Proposal has merit but not realistic and practical	Proposal not yet realistic
Proposal supported by data (case studies, example, statistics, etc.); misconceptions cleared up	Proposal weakened due to lack of supporting data or a few misconceptions evident or both	Proposal not supported by data
Resources accurately documented in given format	Resources documented—format not followed	Resources not yet accurately documented
Displays an understanding of the ethical question and comes up with a believable presentation of both sides	Both sides of the proposal presented	Both sides of the proposal not yet evident
Presents a variety of choices for possible ideas or solutions	Possible ideas or solutions not fully explained	Has not yet presented other possible solutions
Explains the economic impact of the proposal	Presents the economic impact of the proposal	Economic impact not yet presented
Explains how this proposal may affect future generations	Presents how this proposal may affect future generations	Has not yet presented how proposal affects future generations

Note: Realistic = persuasive, well grounded, and documented; practical = acceptable by society and has potential for implementation.

Rubric for Use in a Scientific Investigation[4]

Distinguished Performance	Experienced Performance	Average Performance	Novice Performance
Student can apply all the skills needed to investigate a self-selected issue or problem.	Student can use appropriate skills needed to investigate a self-selected issue or problem.	Student can use some investigative skills to a problem identified by another person.	Student has difficulty identifying skills needed to study a provided problem.
Student can recognize the phenomenon being investigated and develop an explanation for it.	Student can identify the phenomenon being searched for and understands its nature.	Student has difficulty identifying phenomenon being searched for and cannot explain the findings.	Student cannot identify phenomenon being researched and is unable to explain why there is difficulty.

Rubric for Use in a Scientific Investigation[4] *(continued)*

Distinguished Performance	Experienced Performance	Average Performance	Novice Performance
Student can report findings to fellow classmates and to other interested parties	Student can report findings to fellow classmates and make findings available to others	Students can develop a report to give to fellow classmates but has difficulty communicating to others	Student cannot report to fellow students or others the reasons for lack of identification of phenomenon
Student can generalize findings to new and unique situations, thereby using the findings as a tool for investigation	Student can generalize findings to other similar situations and makes appropriate connections	Student can compare findings to other similar situations but has difficulty making connections	Student is unable to recognize similar situations and cannot make connections

The following rubric was designed by a teacher to assess a student's role in a presentation. Expectations for references and presentation format should be communicated to students in advance of the presentation.

Student Role in Group Presentation

3	2	1
Explained and demonstrated very thoroughly his or her role in the presentation	Somewhat explained his or her role in the presentation; role not clearly understood	Said very little if anything about role he or she played in the presentation
Responsibility for his or her part of project was very obvious	Evidence of some time and thought put into part	Minimum of work was done—just enough to meet requirements
Shared ideas and feelings willingly in presentation	Some sharing occurred	Very little if any sharing
Background research very evident—four or more sources mentioned	Some background research evident—two or three sources	Very little evidence of research sources—none to one source
Visuals used were pertinent and easily read	Visuals used but were not easy to understand	Used no media in presentation
Comments for student:		

A science project developed around an issue was the context for the following rubric. In the last category, space could be left so that the actual problems with the project could be written in and feedback provided to the student. Also, in this rubric, content has the same weighting as other components, and if content is of major importance, this area should perhaps receive more weight in the assessment.

Science Issue Project Rubric

3	2	1
Sources ♦ Lists more than three sources ♦ Includes title, page, date ♦ Includes author	Sources ♦ Lists three sources ♦ Includes title, page, date ♦ Includes author	Sources ♦ Lists two sources ♦ Includes title, page, date
Report title ♦ Relates to main topic ♦ Capitalized correctly ♦ Captures reader's interest	Report title ♦ Relates to main topic ♦ Capitalized correctly	Report title ♦ Related to main topic
Content ♦ Many related topics ♦ Informative ♦ Well organized	Content ♦ Relates to topic ♦ Informative	Content ♦ Relates to topic
Paragraphs ♦ Complete sentences ♦ Correct spelling ♦ Correct punctuation ♦ Correct capitalization ♦ Neatly done	Sentences ♦ Complete sentences ♦ Correct spelling ♦ Correct punctuation ♦ Correct capitalization ♦ Neatly done	Sentences ♦ Complete sentences
Illustrations ♦ More than one—titled ♦ Labeled correctly ♦ Neat appearance	Illustrations ♦ Titled ♦ Labeled correctly ♦ Neat appearance	Illustrations ♦ Titled ♦ Labels
Activities ♦ Relates to issue question ♦ Informative ♦ Written conclusion	Activities ♦ Relates to issue question ♦ Informative ♦ Written conclusion	Activities ♦ Relates to issue question
Cover ♦ Title ♦ Name ♦ Date ♦ Illustration	Cover ♦ Title ♦ Name ♦ Date	Cover ♦ Title ♦ Name

Note: All reports will be revised until an A, B, or C is earned. No grades D or F will be awarded.

Peer Teaching: Group Rubric for Issue Presentation

3	2	1
Presentation clear and informative and peer teaching effective; preparation very evident.	Presentation explained but lacked insight and enough preparation for teaching	Presentation not clearly explained; lack of preparation noted
Good selection of items used to support presentation; very interesting, well organized	Selected items were appropriate but presenters did not always seem interested in teaching	Selection of items presented not interesting and little effort shown in preparation
Issue clearly explained and listener gains insight	Issue explained well enough for listener to gain some insight	Issue not explained clearly enough to give listener insight
Sharing and cooperation of group members very obvious	Some sharing and cooperation noted	Sharing and cooperation not obvious
Displayed evidence of having used three or more different kinds of technology	Displayed evidence of using two kinds of technology	One or fewer kinds of technology used
Closing included information and supported conclusions	Closing stated but conclusion did not have much information to support it	Conclusions were not stated
Wrote and sent a thank you to resource people		
Comments:		Total Points _____

Rubric for Model Building to Illustrate Understanding of Electricity Concepts

3	2	1	0
♦ Item of excellent design ♦ Economical and appropriate use of materials ♦ Complete circuit	♦ Item of good design ♦ Some economical and appropriate use of materials ♦ Complete circuit	♦ Item poorly designed ♦ Wasteful or inappropriate use of materials ♦ No complete circuit	♦ No item designed
♦ Picture drawn and completely labeled	♦ Picture drawn but not labeled	♦ Picture drawn but not labeled	♦ No picture drawn
♦ Materials list complete and practical	♦ Materials list complete but not practical	♦ Materials list incomplete	♦ No materials list
♦ Model built and works well ♦ Model neatly done and attractive	♦ Model built but doesn't fully work ♦ Model lacks some attention to detail	♦ Model built but doesn't work ♦ Model is sloppy	♦ No model built
Comments:			

Rubric for Project to Design an Animal Theme Book

Group Report on Book With an Animal Theme		
Great Work	*Acceptable Work*	*Less Acceptable Work*
Tells about the book; includes most significant details without repeating word for word; listener would be able to retell with meaning	Able to relate some significant details; left out some important details that would help others understand the story better; listener may not be able to completely follow story line	Tells some things about the story but details are unrelated or may be insignificant; listener would not be able to repeat the story
Able to rewrite information in own words appropriate to the theme of the book	Able to rewrite information in own words but left out some details important to the story line	Unable to rewrite; copies from original work; merely repeats what was already written
Able to draw illustrations of animals; reflects content of the book's theme	Able to draw illustrations of animals; may not all reflect the book's theme	Unable to complete illustrations or illustrations do not support book theme
Group Proposal for New Book With an Animal Theme		
All took active part in decision making; followed rules and showed courtesy toward others' ideas and actions	Some took active part in decision making; some disruption in the group	Members did not participate in decision making; did not use time to develop proposal
Logical suggestions made for animals to be included in new book	Suggestions needed more clarification but developed a proposal	Some attempt made to develop a proposal but didn't complete task; work needed in group dynamics

Comments:

📖 *A Rubric for Machine Design*

✎ *Task Description*

Imagine that you work for a company that develops machines that can make everyday chores a little easier. You have just been asked to design and make a machine that combines two or more simple machines together to complete one job. You will be expected to complete a drawing of your machine and present at least a 1-minute commercial demonstrating your invention.

Rubric for Machine Design

Excellent	*Good*	*Improvements Needed*
Two or more simple machines used in combination to perform a job	Only one simple machine was used, or used two simple machines but not clear how the two worked together	Simple machines were not used, or student did not appear to know how to complete the task
Invention appears to function in the desired way; realistic proposal	Invention works as shown but not very realistic	Invention does not perform desired function
Simple machines described in a way to show energy conservation (savings)	Simple machines not described well enough to consider energy savings	No description of energy savings
Drawing completed; clear, neat, labeled, detailed	Drawing completed but lacks some details	Drawing completed but quality needs to improve; no attempt made
Commercial clever and creative; well planned	Commercial lacks some creativity; some planning but a few revisions needed	Commercial attempted but more planning needed; no attempt made

Rubric for Construction Plan of an Ice-Fishing Shelter

Task Description

A third-grade class has studied about the four seasons and how the season affects personal activities. During the winter, some students would like to go ice fishing. Richard has agreed to build a shelter for them, but the students must develop a plan for what they want built. Mrs. Morrison insists the shelter be as warm as possible because she gets really cold in the winter. Please do not include anything that may melt the ice because Mrs. Morrison cannot swim and does not want to learn how in cold water.

Draw and label a plan that Richard can follow to build this shelter. Include instructions about materials that are to be used and the dimensions that are to be used. Last, include a paragraph explaining why your plan is the one our class should present to Richard. Give reasons that will convince us that you have developed the best plan.

Rubric for Construction Plan of an Ice-Fishing Shelter

3	2	1
Good selection of materials. They will do the job and provide insulation from the cold, reasonable comfort and durability.	Information about materials is provided but may not be the best materials for the job.	Does not provide information about materials used.
Dimensions are complete and reasonable. A good shelter should result.	Dimensions are given. May be unreasonable for the job or may be incomplete.	Doesn't give dimensions for the shelter.
Paragraph contains sufficient and good reasons why your plan is best. Should be accepted by the builder.	Paragraph explaining why plan is best is provided. May not be backed up with facts.	Paragraph describing why your plan is best is missing or superficial.
Project is clearly written and easy to understand. Attractive presentation causes builder to choose your design.	Project can be understood but lacks attractive presentation.	Project is messy and difficult to understand.
No errors in computation to interfere with successful completion of the shelter.	Enough errors in computation to cause finished product to turn out incorrectly.	Major errors have been made in computations.

📖 *How Well Does a Rubric Communicate the Original Task?*

The following rubric was designed to assess a wildlife project that students were to complete. The rubric should communicate the major expectations for the tasks that students are to perform, and from the rubric, the reader should at least be able to construct the framework of the original task. The rubric should also send the message of what is valued in the performance. Can the original task be constructed from this rubric?

Criteria for A Wildlife Project

Criteria	5	3	1
Appropriate season and climate	States in which seasons your projects will be used, including why they are appropriate	Seasons stated but appropriateness not stated	Invalid seasons and climate
Appropriate plants and animals for biome	Project appropriate for the animals and plants in your area	Project appropriate for either animals or plants in your area	Project is invalid for animals and plants in your area

Criteria for A Wildlife Project *(continued)*

Criteria	5	3	1
Description of biome food web	Describes how your project benefits your area, including a six- to nine-organism food web	Description of benefits and four- to five-organism food web	Description of benefits and less than four-organism food web
Description of environmental maintenance	Description of how your project will maintain the balance of nature	Description only of your organism, not of surrounding environment	Description invalid for organism or environment
Presence of basic needs of organisms	Project adequately supports at least one basic need of organisms	Project inadequately supports the basic needs	Project does not support a basic need

Bottom Line: Not Every Rubric 📖 *Is a Good or Useful Rubric*

The following rubric is a work in progress, in the "not yet acceptable" category. Consider changing some of the wording. Does the word *describes* convey something different than the word *explains?* Are aesthetics important? The rubric conveys that aesthetics are valued, but what is beautiful to the teacher may not be so to the student. Other categories need work as well. What is *cramped?* The purpose of including this example is that it is a start, but it serves as a good example of a rubric that is not very useful for assessing a portfolio. If your classroom uses rubrics, revisit your goals and objectives, and structure your rubric around these.

Portfolio Assessment Rubric: An Example of a Not Very Useful Rubric

Criteria	4	3	2	1
Explains what my successes are	Clear, complete, some creativity to clarify	Clear, complete	Complete but unclear	Sketchy, vague
Explains what is presently being done in my classes	Gives specifics and clearly identifies major parts of each unit	Gives specifics of all units or projects	Gives some specifics	Only vague ideas
Pleasant to look at	Beautiful	Nice	Average	Plain
Easy to read	Clear, correct, fun to read	Clear and correct	Clear	Fairly clear
Organized	Well laid out with neatness	Neat, no direction	Cramped	Cramped and messy

📖 *A Second Example of a*
Not Very Useful Rubric

The following rubric could be the template for a rubric to assess a laboratory performance related to the effects of gravity, friction, and mass on motion. From the rubric, it appears that this may be some type of culminating assessment, because the effects of three variables are being considered. Some problematic words exist in this rubric. What is meant by *accurately*? What does *relate* or *related* mean? Nothing about the level of conceptual understanding is really conveyed. It sounds as if a textbook definition would also do the job, and that is probably not what the rubric developer had in mind. Rubrics take revision and refining. The Motions Concepts Rubric is in the "not quite there yet" stage.

Motion Concepts Rubric

5	4	3	2	1
Accurately computed speed		Attempted formula but miscalculated		Made measurements and attempted computations
Accurately explained the effect that gravity has on motion		Partial description of gravity		Unable to relate gravity to motion
Accurately explained the effect friction has on motion, gave an example		Gave partial description of friction or gave an example of friction		Unable to relate friction to motion
Accurately explained the effect mass has on motion		Gave partial description of the effect mass has on motion		Unable to relate mass to motion
Accurately related three laws of motion		Accurately related two laws of motion		Accurately related one law of motion

A Third Example of a Not Very Useful Rubric 📖

Suppose students were asked to follow this rubric in preparing an essay on ecology:

Criteria	3	2	1
Cover	Attractive, colorful	Colorful	No color
Essay presents a lesson	Clear and creatively presented	Stated clearly	Not clear
Supports the lesson	Vivid details, understand	Little content	No content
Grammar	Correctly presented	Presented one to four errors	Many errors
Use of capital letters	Correctly presented	One to three errors	Many errors
Illustrations	Clear, colorful	Clear	Not complete

The message conveyed by this rubric seems to be that the essay is to focus on writing mechanics. This seems counter to the title, "Essay on Ecology," which appears to indicate that some criteria about understanding selected ecological concepts might have been expected. If the essay is going to be scored on writing mechanics, a separate score for these elements could be awarded, and a second rubric could be used to assess conceptual understanding. Also, no context, other than ecology, is provided with this rubric.

Evaluation of Science Frameworks

With the focus on classroom assessment, the alignment of the curriculum with district science learning outcomes becomes a task that often involves teachers and other district personnel. A framework rubric, aligned with the NSES, could be used to focus the evaluation process when districts and teachers examine and document the elements of the frameworks that are in place or those being established. The following evaluation form, based on the NSES, was designed by the staff of the Iowa—Scope, Sequence, and Coordination Project in their work with districts engaged in editing their science frameworks.

📖 *Science Framework Alignment With NSES*

   This rubric is based on the NSES (NRC, 1996), and page numbers in parentheses refer to the location in the NSES. Prior to framework alignment, some discussion should be held about what constitutes *evidence*. The descriptors for the level of evidence can be set by the group completing the framework or curriculum evaluation. This example shown in Form 4.1 is provided as a tool not only for evaluation but also for within-district communication about the science curriculum.

NOTES

1. Adapted from Mason City Schools, Mason City, Iowa.
2. Adapted from Mason City Schools, Mason City, Iowa.
3. Adapted from Mason City Schools, Mason City, Iowa.
4. Adapted from Mason City Schools, Mason City, Iowa.

FORM 4.1

1. Rationale

Elements	Not Evident	2	3	4	Evident
1.1 The rationale of the framework is succinctly written.					
1.2 The rationale explains how district goals align with the framework.					
1.3 The rationale explains how Science-Technology-Society is infused into the framework.					
1.4 The rationale explains how constructivism is incorporated into the framework.					
1.5 The rationale describes how the district's tenets fit into the framework.					
1.6 The rationale explains how National Science Education Standards fit into the framework.					
1.7 The rationale is written so that it is meaningful to a diverse audience.					

Comments:

2. Evidence of Science Literacy

Elements	Not Evident	2	3	4	Evident
2.1 The framework reflects a richness of experience of knowing about and understanding the natural world (p. 13).					
2.2 The framework includes scientific processes and principles that can be used to make personal decisions (p. 13).					
2.3 The framework reflects an expectation that students will engage intelligently in public discourse and debate about matters of scientific and technological concern in the classroom (p. 13).					

Comments:

FORM 4.1 Continued

3. Content Standards, Grades 5–12—Unifying Themes (p. 115)

3.1 The *broad unifying themes and processes* are evident at all grade levels. The broad unifying themes and processes complement the analytic, more discipline-based perspectives presented in the other content standards	Not Evident	2	3	4
3.1a Systems, order, and organization (p. 116)				
3.1b Evidence, models, and explanation (p. 117)				
3.1c Change, constancy, and measurement (p 117)				
3.1d Evolution, and equilibrium (p. 119)				
3.1e Form and function (p. 119)				
Comments:				

3. Content Standards, Grades 5-12—Inquiry

3.2 The framework reflects an inquiry-based approach to instruction at all grade levels. Inquiries are based on the types of questions explored that influence methodology, techniques, and core theories (p. 143).	Not Evident	2	3	4	5
3.2a The framework includes activities that develop student abilities necessary for conducting scientific inquiry, such as refining and refocusing broad and ill-defined questions; enhancing process skills; using appropriate tools; and thinking critically about evidence, anomalies, and so forth (pp. 110, 145).					
3.2b The framework includes activities that develop student understanding about the process of scientific inquiry. Inquiries should lead not only to greater understandings but also lead to new investigations (pp. 110, 148).					
Comments:					

FORM 4.1 Continued

4. Content Standards, Grades 5–8—Physical, Life, and Earth and Space Science; Science and Technology; Science in Personal and Social Perspectives; and History and Nature of Science

4.1 The *physical science* standards for Grades 5–8 are evidenced through student experiences based on the content, process, and concept goals outlined in the framework (pp. 106, 108).	*Not Evident*	*2*	*3*	*4*	*Evident*
4.1a Properties and changes in matter					
4.1b Motions and forces					
4.1c Transfer of energy					
4.2 The *life science* standards for Grades 5–8 are evidenced through student experiences based on the content, process, and concept goals outlined in the framework (pp. 106, 108).	*Not Evident*	*2*	*3*	*4*	*Evident*
4.2a Structure and function in living systems					
4.2b Reproduction and heredity					
4.2c Regulation and behavior					
4.2d Populations and ecosystems					
4.2e Diversity and adaptations of organisms					
4.3 The *Earth and space science* standards for Grades 5–8 are evidenced through student experiences based on the content, process, and concept goals outlined in the framework (pp. 106, 108).	*Not Evident*	*2*	*3*	*4*	*Evident*
4.3a Structure of the earth systems					
4.3b Earth's history					
4.3c Earth in the solar system					
4.4 *Science and technology* student experiences for Grades 5–8 explore the connections between the natural and designed world. These are based on the content, process, and concept goals outlined in the framework (pp. 106, 108, 161).	*Not Evident*	*2*	*3*	*4*	*Evident*
4.4a Abilities of technological design to meet human needs, solve problems, develop a product, and so on					
4.4b Understanding the relationship between science and technology—distinctions, similarities, roles, and relationships					

FORM 4.1 Continued

4. Content Standards, Grades 5–8 *(continued)*

	Not Evident	2	3	4	Evident
4.5 Through the linkage of *science in personal and social perspectives*, students in Grades 5–8 will have the opportunity to develop a means to understand and act on personal and societal issues. These are based on the content, process, and concept goals outlined in the framework (pp. 107, 108, 166-170).	Not Evident	2	3	4	Evident
4.5a Personal and community health					
4.5b Population growth					
4.5c Natural resources					
4.5d Environmental quality					
4.5e Natural and human-induced hazards					
4.5f Science and technology in local, national, and global perspectives					
4.6 *History and nature of science* experiences are interwoven throughout the study of science for Grades 5–8. These are based on the content, process, and concept goals outlined in the framework (pp. 107-108, 170-171). These are evidenced through experience in three broad areas.	Not Evident	2	3	4	Evident
4.6a Science as a human endeavor—relying on insights, skill, creativity, scientific habits of mind, tolerance of ambiguity, skepticism, and openness to new ideas					
4.6b Nature of science—tentative nature of knowledge; experimental and observation evidence that contributes to major, established ideas of science; methods of scientific inquiry					
4.6c History of science—illuminating the history of innovation and thinking, practices in different cultures, and contributions of individual women and men					

Comments:

FORM 4.1 Continued

5. Content Standards, Grades 9–12—Physical, Life, and Earth and Space Science; Science and Technology; Science in Personal and Social Perspectives; and History and Nature of Science

5.1 The *physical science* standards for Grades 9–12 are evidenced through student experiences based on the content, process, and concept goals outlined in the framework (pp. 106, 108).	Not Evident	2	3	4	Evident
5.1a Structure of atoms					
5.1b Structure and properties of matter					
5.1c Chemical reactions					
5.1d Motion and forces					
5.1e Conservation of energy and increase in disorder					
5.1f Interactions of energy and matter					
5.2 The *life science* standards for Grades 9–12 are evidenced through student experiences based on the content, process, and concept goals outlined in the framework (pp. 106, 108).	Not Evident	2	3	4	Evident
5.2a The cell					
5.2b Molecular basis of heredity					
5.2c Biological evolution					
5.2d Interdependence of organisms					
5.2e Regulation and behavior					
5.2f Matter, energy, and organization in living systems					
5.2g Behavior of organisms					
5.3 The *Earth and space science* standards for Grades 9–12 are evidenced through student experiences based on the content, process, concept goals outlined in the framework (pp. 106, 108).	Not Evident	2	3	4	Evident
5.3a Energy in the earth system					
5.3b Geochemical cycles					
5.3c Origin and evolution of the earth system					
5.3d Origin and evolution of the universe					

FORM 4.1 Continued

5. Content Standards, Grades 9–12 *(continued)*

5.4 *Science and technology* student experiences for Grades 9–12 explore the connections between the natural and designed world. These are based on the content, process, and concept goals outlined in the framework (pp. 107, 108, 161)	*Not Evident*	*2*	*3*	*4*	*Evident*
5.4a Abilities of technological design to meet human needs, solve problems, develop a product, and so on					
5.4b Understandings about science and technologies—distinctions, similarities, roles, and relationships					
5.5 Through the linkage of *science in personal and social perspectives,* students in Grades 9–12, will have the opportunity to develop a means to understand and act on personal and societal issues. These are based on the content, process, and concept goals in the framework (pp. 107, 108).	*Not Evident*	*2*	*3*	*4*	*Evident*
5.5a Personal and community health					
5.5b Population growth					
5.5c Natural resources					
5.5d Environmental quality					
5.5e Natural and human induced hazards					
5.5f Science and technology in local, national, and global perspectives					

Comments:

FORM 4.1 Continued

5. Content Standards, Grades 9–12 *(continued)*

	Not Evident	2	3	4	Evident
5.6 *History and nature of science* experiences are interwoven throughout the study of science for Grades 9–12. These are based on the content, process, and concept goals outlined in the frameworks (pp. 107-108, 170-171). These are evidenced through experience in three broad areas.					
5.6a Science as a human endeavor—ethical and moral traditions and dilemmas, influence of society and culture, the range and variety of scientific endeavor					
5.6b Nature of scientific knowledge—the empirical standards that differentiate science from other ways of knowing; criteria for scientific explanations, such as consistency, repeatability, and usefulness in making predictions					
5.6c Historical perspectives—practices of different cultures; contributions of individuals in their daily work; building on earlier knowledge contrasted with occasional and historic leaps of understanding; noteworthy advances, such as germ theory and plate tectonics					

Comments:

6. Content Standards, General
This section represents a general evaluation of the Grades 5–12 content areas based on the review of the parts.

Elements	Not Evident	2	3	4	Evident
6.1 The concepts and processes found in the framework show connections among the scientific disciplines (p. 115)					
6.2 The concepts and processes found in the framework are at developmentally appropriate grade levels (p. 115).					
6.3 The continuum of complexity of the *sequence* (i.e., concrete to abstract) is evident in the framework (pp. 115-116).					
6.4 The framework includes integrative schemes, typified by the unifying concepts and processes (p. 116).					

Comments:

FORM 4.1 Continued

7. Assessment Standards, Grades 5–12

Elements	Not Evident	2	3	4	Evident
7.1 The framework explains the purpose of the assessments used (p. 78).					
7.2 Examples of assessments that assess achievement and are used for grading are included at all levels in the framework (p. 79).					
7.3 Assessments that probe for understanding, higher-order reasoning, and the application and use of knowledge are included at all levels in the framework (p. 82).					
7.4. Assessments at all levels are based on the content standards (p. 79).					
7.5. Assessment examples included in the framework reflect the nature of the classroom learning environment (i.e., they are based on and tested in the context of study, similar to activities of scientists, and can be applied to everyday experiences (p. 83).					
7.6 Assessments at all levels are fair. They strive to be developmentally appropriate, use a familiar setting, are unbiased; students are given adequate time to demonstrate their achievement (pp. 83-4).					
7.7 Assessment examples in the framework are issue based, set in a variety of contexts, and show personal relevance for students (p. 83).					
7.8 Assessment examples in the framework include those used for the purposes of determining students' initial understandings and abilities (p. 87).					
7.9 Assessment examples included in the framework are diverse, i.e., include opportunities for students to explain orally, in writing, or through illustration how a work sample provides evidence of understanding (p. 88).					
7.10 Assessment examples included in the framework provide students with opportunities to evaluate and reflect on their own scientific understanding and abilities (p. 88).					

Comments:

SOURCE: © 1997 Iowa Scope, Sequence, and Coordination Staff. Reprinted with permission.

5

Assessment Examples for All Grade Levels

In this chapter, you can find suggestions for the following:

- Opportunities for student self-assessment in group work

- Assessing students' resource use for projects or assignments

- Using open-ended questions

- Using the Student Laboratory Environment Inventory (SLEI)

- Assessing in the application domain

- Using an observation checklist

- Using activities to foster creativity

- Using application-level multiple-choice questions

- Encouraging analogical and metaphorical thinking

- Assessing attitudes, preferences, and processes, including the STS Attitude Scale

Having students assess their own personal work and their performance in a group can assist them in viewing their roles in the accomplishment of group tasks. Several suggestions for student self-assessment in small-group work are presented in Forms 5.1, 5.2, and 5.3.

An example of reliance on students' self-evaluation is the resource use inventory; an example is shown in Form 5.3. This inventory can help

FORM 5.1

Student Self-Assessment:
Group Evaluation (Attitude Domain)

Type of Group Activity: _____

Student Name (first and last): _____

For questions 1 and 2, circle the words that describe how you feel.

1. How well did your group work together?	*Very well*	*Well*	*Not very well*
2. Overall, how would you rate your group's product (poster, pamphlet, skit, etc.)?	*Very good*	*Good*	*Not very good*

3. What suggestions do you have for helping groups work together?

_____

_____

_____

4. What do you think was the best part of your group's product?

_____

_____

_____

5. How do you think your group could have improved their product?

_____

_____

_____

FORM 5.2

Individual Student Assessment (Attitude Domain)

Name: _____

For questions 1 through 3, circle the words that describe how you think.

1. How much did you contribute to the group product?	*More than others*	*Same as others*	*Less than others*
2. Did you offer ideas?	*More than others*	*Same as others*	*Less than others*
3. Did you accept ideas from the group?	*More than others*	*Same as others*	*Less than others*

4. What would you like others to know about the work that you did on this product?

_____

_____

_____

FORM 5.3

Resources Used in Class-Related Work (Application Domain)

Read each of the following questions, and then check the column that best describes your use of the information resources. If you check Column 2, estimate the number of times and enter the number in Column 3. If never used, check the last column.

Resource Use	Once in the Past 3 Months	More Than Once in the Past 3 Months	Estimated Number of Times	Never Used
1. Have you used a phone to gather information for class work?				
2. Have you used materials from the school library or media center for class work?				
3. Have you used materials from other sources for class work?				
4. Have you used a computer for class work?				
5. Did you discuss science topics with adults?				
6. Did you use the Internet to help with class work?				
7. Did you use the TV or radio to help with class work?				
8. Did you use newspapers, magazines, or journals to help with class work?				
9. Did you take notes to help you remember?				
10. Did you take notes when doing research?				
11. Did you take notes without being told to do so?				
12. Have you contacted resource people to find information?				
13. Have you contacted a speaker to supplement a research topic?				
14. Have you introduced a resource person to the class before she or he makes a presentation?				
15. Has interviewing been a part of your information gathering for work related to class?				

Comments:

provide focus for work in which student use of information beyond the textbook is desired. Making connections via accessing a wider range of materials can encourage greater student independence in learning.

Open-Ended Questions (Integrates Domains)

Students can respond to open-ended questions by providing solutions to problems, explanations, and opinions. The responses can reveal students' ideas and conceptions much like an individual oral interview (Strauss & Stavey, 1983). A variant of open-ended questions requires students to provide definitions. The interpretation of responses does then require some caution because students can memorize definitions by rote (Ausubel, 1968). However, incorrect definitions sometimes can reveal misconceptions (Barenholz & Tamir, 1992). In the following example, a statement is given, a question follows, and students are asked to explain.

Example: People are able to live healthy lives eating only fruits and vegetables and no meat. Explain how this can be possible.

Possible Explanation: No one fruit, vegetable, or meat contains all the necessary nutrients for healthy living. A vegetarian diet can be eaten that would provide the nutrients essential for good health. A vegetarian would have to be certain to include nonmeat foods that provide any of the proteins supplied by meat.

Questions for Students to Address 📖

Students could first be given the question of how the caloric intake of a vegetarian compares with that of a nonvegetarian. To further initiate their thinking about caloric intake, students could be asked to respond to the following questions:

- How do the numbers of calories in 10 grams of bread, lean hamburger, and celery compare?

- Draw a graph showing the number of calories available in 10 grams each of bread, lean hamburger, and celery.

Performance Expectation: The students should draw a graph, provide an appropriate title, and record the units. The results are likely to show that the caloric values of meat and bread are very similar, whereas celery has a much lower caloric value.

- What would you need to know to determine if school lunches were nutritious and balanced?

Performance Expectation: School lunch menus, portion sizes, and food preparation details would need to be analyzed. The chemical composition of the lunches, the fat, carbohydrate, and protein content, their caloric values, and the relative amounts of nutrients would need to be considered in the answer.

Student Laboratory Environment Inventory (Integrates Domains)

Although much is known about hands-on learning, and many studies have corroborated that learning occurs best when actually doing science, the research has not been comprehensive in assessing the effects of laboratory instruction on student learning and attitudes. Laboratory experience is an expectation in the sciences, and the Student Laboratory Environment Inventory (SLEI; Fraser, Giddings, & McRobbie, 1992) was designed to assess student attitudes about learning in the science laboratory. The SLEI is intended for use where a separate laboratory class exists, and this instrument may be more appropriate for upper secondary and higher education levels. The 35 items, with response alternatives of *almost never, seldom, sometimes, often,* and *very often,* measure five dimensions. Two forms of the instrument have been designed. An Actual Form asks students to note what actually takes place in the laboratory classroom, and the Preferred Form asks students to respond to what they would prefer to take place in the laboratory class.

Each of the aforementioned assessment activities can be completed in the classroom within the realm of authentic assessment that is embedded in the instruction. The students may be performing any one of these process skills during an instructional period, and the teacher could, using a predetermined observation checklist, observe the groups performing a particular class activity to determine a student's ability to perform that process. Form 5.4 includes examples of items that could be used to assess some of the processes of science.

Ideas for Assessing in the Application Domain

Assessing student learning in their abilities to make applications involves asking them to apply principles or skills (or both) to new situations for which they are expected to provide unique solutions. Prior to assessing a student's ability to apply concepts, teachers should develop

FORM 5.4

Checking Process Skills

Student _____ Date _____

Activity _____ Process (skills) _____

(Strongly Agree = SA, Agree = A, No Opinion = N, Disagree = D, Strongly Disagree = SD)

1. This student was using appropriate process skills to determine the outcome of the activity.
   SA   A   N   D   SD

2. This student performed the process skill within the context of the activity.
   SA   A   N   D   SD

3. The process skill enhanced the learning taking place during this activity.
   SA   A   N   D   SD

4. The other student performing this activity (agreed, no opinion, disagreed) with the performed process skill.
   SA   A   N   D   SD

5. This student noted appropriately the performance of the expected process skill in the notebook entry.
   SA   A   N   D   SD

6. This student needs additional experiences to practice this _____ skill.
   Yes   No

Comments:

Enger, S., & Yager, R. *Assessing Student Understanding in Science.* © 2001 by Corwin Press, Inc.

questions that promote the thinking involved in applications. Certain words and phrases can frame the thinking required in making applications.

Possible Prompts	Application Words	Application Words
What would you use to ..	Separate	Predict
What is the use for . . .	Arrange	Identify, classify
What would result if . . .	Select	Explain
Tell what would happen if . . .	Differentiate	Construct
Tell how much change there might be if . . .	Demonstrate	Plan
Illustrate how . . .	Show how	Conclude
Choose the statements that apply . . .	Cause for	Diagram
What is the reason for . . .	Compare and contrast	Outline
	Investigate	Chart
	Dissect	Graph

Some examples of application questions follow, and these questions typically would be followed with further questions related to the how and why. Addressing the how and why can move students to using higher-order thinking skills.

Which of the following methods is best for _____?

What steps should be followed in applying _____?

Which situation would require the use of _____?

Which principle would be best for solving _____?

What procedure is best for improving _____?

What procedure is best for constructing _____?

Which of the following is the best plan for _____?

What is the most probable effect of_____?

Application questions that call for information use or skills often require original thinking because these questions may

- Require knowledge of standard procedures

- Present a specific situation and require appropriate action, reasonable inference, or prediction

- Require solutions of numerical or other mathematical problems

- Present a statement or observation to be interpreted or evaluated

- Assume unusual or complex situations and require inferences

When constructing assessment instruments or items, the words and phrases used indicate the expected level of cognitive processing. The students should have the chance to integrate their procedural understanding as well as their conceptual learning to be able to respond to the questions. If the questions are designed to only assess student recall abilities, these questions require less cognitive processing. Questions are desirable that ask students to combine processes and concepts they have been learning and then make applications. Students should have opportunities to use higher-order thinking skills, and questioning strategies can promote this construct.

Observation Checklist (Integrates Domains)

The following example of assessment has been embedded within the instructional phase of a unit or module addressing such issues as flooding, ecosystems, wetlands, water pollution, weather, and recreation.

Example: The students are given an activity in which two pertinent questions are asked. The students should form cooperative groups and be asked to investigate the questions provided in the activity. They may use any method on which they agree, and the solution must be in a form that can be presented to the whole class. The time frame for this activity can be decided on either by the teacher or by a vote by the students.

While the students are working on this investigation, the teacher may want to perform an embedded observational assessment provided in the following table:

Observed Behaviors	What Was Observed	Notes, Comments
Student groups are on task.		
Students are actively discussing issue.		
Students are using prior concepts to solve problem.		
Students are exhibiting a positive attitude.		
Students are demonstrating an understanding of scientific problem solving.		
Students are using multiple strategies to address the problem.		
Students are using a variety of skills to present a group solution.		
Other		

This kind of checklist could be electronically supported or placed on an index card and carried by the teacher while circulating among the cooperative groups. Assessable activities may include (a) students talking about their own experiences, thereby bringing prior knowledge into play (concepts may be overheard and shared by students); (b) communication skills within groups; (c) attitudes of individual group members or entire groups; (d) newly generated questions; or (e) creativity dealing with innumerable variables. There are many more observations that can be noted depending on the prior criteria set up by the teacher and students.

◆ Use of Application-Level Multiple-Choice Questions

Multiple-choice questions can be written to assess concepts at the application level, and a search of test publishers' materials and state assessment documents should also provide a source of these kinds of questions. Examples are given in Figure 5.1; for each question, the best answer is indicated by as asterisk.

FIGURE 5.1. Sample Application Assessment Items (Application Domain)

The best answers are indicated by an asterisk.

1. Concept: When water freezes, its volume increases.

   Application Question: Which one of the following is the main reason that water should not be stored in the freezer in a container totally filled and sealed?

   A. The taste of the water will change.
   B.* The container might break as the water expands.
   C. The water reacts with the glass at low temperatures.
   D. Water will not freeze if there is not enough space for it to change to ice.

2. Concept: The time to warm an object that is in a boiling liquid depends on the amount of material making up the object and how much of its surface is exposed to the boiling liquid.

   Application Question: Which of the following examples will cook most slowly when placed in boiling water?

   A.* A single 1-pound potato
   B. One pound of small potatoes
   C. One pound of medium-sized potatoes
   D. One pound of potatoes cut into small pieces

3. Concept: Most bird identification guides are based on knowledge of a bird's shape, size, color, and patterns of markings.

   Application Question: While sitting at the breakfast table on a winter morning, you notice a species of bird you have never seen before. What one of the following would you recommend to best guarantee that you will be able to identify the bird?

   A. Make a note of the bird's favorite food.
   B. Observe the behavior of the bird.
   C.* Study the size and coloration of the bird.
   D. Determine the sex of the bird.

4. Concept: The temperature at which water boils decreases with altitude; therefore, the temperature of boiling water at high altitudes will not be as high as the temperature of boiling water at sea level. A pressure cooker is a kitchen appliance where high pressure and high temperature are maintained inside the cooker despite the altitude.

   Application Question: Where would it be more efficient to have a pressure cooker for cooking food?

   A. Below sea level
   B. At sea level
   C. At low elevations
   D.* In the high mountains

(continued)

FIGURE 5.1. Sample Application Assessment Items *(continued)*

5. Concept: A great amount of heat energy is required to evaporate water; therefore, evaporation is used as cooling process.

   Application Question: If you were out on a camping trip, which of the following situations would result in providing the coldest drinking water? Assume that the amount and beginning water temperature is the same for each case.

   A. A metal canteen is filled with water and kept in the shade.
   B.* A metal canteen is filled with water, covered with a wet cloth, and then kept in the shade.
   C. A metal canteen is filled with water, immersed in a bucket of water at the same temperature as inside the canteen, and then kept in the shade.
   D. A metal canteen is filled with water and kept in direct sunlight.

6. Concept: Light-colored objects reflect sunlight better than dark-colored objects.

   Application Question: On a sunny winter day, which vehicle would be the warmest to the touch?

   A. A blue car
   B. A red car
   C. A white car
   D.* A black car

7. Concept: In very cold conditions, up to 80% of the heat made by the body can be lost through the surface of the head and neck.

   Application Question: If you were suddenly caught outside in below-zero weather and all you had to wear were boots, shorts, a T-shirt, and a light jacket, which of the following would be the best way to retain the greatest amount of body heat?

   A. Wrap the T-shirt around your bare legs.
   B. Leave the T-shirt on as you usually wore it.
   C.* Wrap the T-shirt around your head and neck.
   D. Use the T-shirt to keep the nearby air moving.

8. Concept: Objects with a large ratio of surface area to volume will cool faster than objects with a small ratio of surface area to volume.

   Application Question: Suppose a waiter brings you cooked steak. Which is the best way to keep the steak as warm as possible while you eat it?

   A.* Cut only the piece to be eaten
   B. Cut the steak quickly into bite-sized pieces
   C. Keep air moving near the steak
   D. Eat slowly

9. Concept: Warm-blooded animals lose body heat in proportion to their surface area and generate body heat in proportion to their body mass.

   Application Question: Which of the following animals would need to eat more per gram of body mass to maintain body heat?

   A.* A mouse.
   B. A dog.
   C. A cow.
   D. A cat.

FIGURE 5.1. *(continued)*

10. Concept: When most metal objects are heated, they increase in size.

   Application Question: What assumption are you making when you run hot water over a metal lid on a glass jar to loosen the lid?

   A.   Both the glass jar and metal lid increase in size in the same proportion.
   B.   The glass increases in size in a greater proportion than the metal lid.
   C.* The metal lid increases in size in a greater proportion than the glass jar.
   D.   Glass and metal do not stick together as much in water.

11. Concept: When most metal objects are heated, they increase in size.

   Application Question: If a nickel with a hole in it is heated, what will happen to the size of the hole?

   A.   The hole will decrease in size.
   B.   The hole will stay the same size.
   C.* The hole will increase in size.
   D.   The hole will become irregular.

12. Concept: Steam can be at the same temperature as boiling water, but steam has the energy of vaporization that is released when it condenses to water.

   Application Question: Which of the following burns will tend to be most damaging to the skin?

   A.   Burns caused by boiling water
   B.* Burns caused by steam
   C.   Burns caused by water vapor
   D.   Burns caused by hot tap water

13. Concept: The color of an object is determined by the wavelengths of light that it reflects and those it absorbs.

   Application Question: Which of the following best explains why most plants are green?

   A.* Green light is reflected from the plants.
   B.   Green light is absorbed by the plants.
   C.   Green light is not needed by the plants.
   D.   Green light is harmful to plants.

Examples for the Creativity Domain

Fostering Creativity Using Creative Situations 📖
(Creativity Domain)

The use of creative writing is a way to foster creativity in students. Conceptual understanding can be monitored with this type of assessment as well. The sample questions in this section can be used to stimulate the imagination and incorporate as many or as few concepts as the teacher deems necessary.

The teacher should set the stage for the students and could introduce the writing activity like this:

I am going ask some questions (or describe some situations) that will give you a chance to see how you are at thinking up new ideas and solving problems. I want you to use all the imagination and thinking ability you have. This is like a game to exercise your brain. You will have a chance to use your imagination in thinking up ideas and putting them into words Try to think of as many ideas as possible. Try to think of interesting, unusual, and clever ideas. It does not matter if you have the same ideas as others in the class or if you suggest something that no one else thinks up.

Here are two sample questions:

Elementary: "What would the world be like if it rained chocolate drops?"

Secondary: "What would happen if lakes became frozen from the bottom?"

Situation statements should be related to the unit of instruction so that the students could connect the assessments and what they are studying. Care should be taken, however, to ensure that the unit of instruction does not center on the situation statement to such an extent that this measure of creativity becomes a test of knowledge.

Teachers may want to present situations related to concepts being studied and have students write their responses.

Sample situation statements could include the following:

First grade: "Bobby woke up and found dinosaurs in his yard."
              What would your reactions be?

Third grade: "Suppose we lived in a world without insects."
              What would your reactions be?

Fifth grade: "Pretend that there was no more pollution."
              What would your reactions be?

Seventh   "Suppose there was no more disease in the world."
  grade: What would your reactions be?

General: "Jane stopped at the gas station to buy gas for her car."
              What would your reactions be?

Scoring This Creativity Task ✎

The rationale behind this measure is to provide a thought-provoking situation appropriate to the ability and experiences of the students to be assessed and to have students write as many pertinent and imaginative responses to the situation as possible. The number of such responses will provide a clue to their overall creativity. These questions can be used to look at creativity by examining two factors: (a) the number of questions asked and statements made by the student and (b) the quality and uniqueness of those questions and statements.

Care must be taken to correctly frame the interpretation of student responses. Suppose the students were asked to respond to the situation statement that "Chris woke up and found dinosaurs in his front yard." A student could say that this occurred because the Earth went through a time warp. This could certainly be considered unique, but if many students said that, the response would simply be pertinent. A cartoon show may have been a common experience that generated this thinking among numerous students. You could only make the determination of frequency by examining all of the student responses.

Student responses could be evaluated by using a coding system based on relevance, pertinence, and uniqueness. Consider the following examples:

Statement: Suppose that you got up one morning and found that there was no gravity.

Sample student response: Dogs would chase cats.

This would be judged irrelevant because the student's response is not related to the question. If another student responded that we would float away, this response would be pertinent but not particularly creative. A unique response might be that he or she would be able to jump very high to pick fruit from trees. After evaluating student responses as irrelevant, pertinent, or unique, the responses in each category would be counted.

A Reminder: Foster Creativity, Don't Stamp It Out ✎

The foregoing kind of activity can be conducted from time to time with the idea of mental stretching. The familiarity of the topic statement would very likely affect the range of responses. Validity and reliability are major concerns when using this kind of assessment for making judgments about student creativity. These can be fun activities to stretch thinking, but at no time should students be labeled on the basis of the results in this kind of assessment activity. The creative spirit should be fostered and not dampened.

📖 *Encouraging Analogical and Metaphorical Thinking (Creativity Domain)*

Encourage analogical and metaphorical thinking in the classroom, and present these opportunities to students. Students should have opportunities to generate analogies and then explain what the analogy supports and what it does not support.

As an example, students could be given analogies about human biology. Students could be asked to first complete correspondences as follows:

The heart is like _____

The brain is like _____

The skeleton is like _____

Students could then be asked to further analyze the validity of the analogy. For the phrase "The brain is like _____," a student response might be, "The brain is like a computer. This comparison works because parts of the brain process information. This comparison is not quite accurate because our brains can think independently, but computers do not really think."

The assessment of these kinds of activities is not a particularly easy task because of the subjectivity involved. This may be the kind of activity on which students are asked to reflect over time about their capabilities. Schools want to foster students who are good thinkers, and analogical and metaphorical thinking support higher-order thinking. These are obviously not easy attributes to assess, but evidence of having these opportunities in the curriculum can be documented.

📖 *Drawing Production as a Creative Thinking Assessment (Creativity Domain)*[1]

Creativity in science may be ignored in the traditional science classroom, but the creative spirit is often an attribute of scientists. Students can be exposed to situations in which they are expected to propose visual solutions, as in the following example. Students might be presented with this scenario:

A scientist was making a drawing and was interrupted before finishing it. You found the paper and are going to help finish the drawing. You may complete the drawing in any way that you like. The scientist will be back in 15 minutes, so your work should be finished in that time.

How do you assess this drawing production (DP)?

DP allows students of most ages and ability groups to interpret and complete what they conceive to be significant for the development of a creative product. This approach has been viewed as an important addition to the culture-fair assessment of creative potential (Jellen & Urban, 1986). Eleven criteria can serve as the evaluation framework to assess creativity in a drawing production; these are interacting entities reflecting a holistic concept to creative thought.

The DP construct is also supported by those components of creative thought that can be found throughout the existing literature on creativity and creativity testing. These components are fluency, flexibility, originality, and elaboration. Three other components of creativity that can also be assessed via this approach are risk taking, composition, and humor as a cognitive-affective ability capable of freeing the mind from concrete or unpleasant realities (Jellen & Urban, 1986).

Jellen and Urban (1986) have suggested the following 11 key elements as evaluation criteria for the Test for Creative Thinking in Drawing Production:

1. Completion (Cm)—Any continuation and extension of the six figure fragments. One point for each continuation. Point value: six points maximum.

2. Additions (Ad)—Any additions made to the extended or continued figure fragments (e.g., points, lines, patterns, etc.). One point for each addition. Point value: six points maximum.

3. New elements (Ne)—Any new figure, symbol or element created in addition to, but independent of, the given figure fragments. One point for each connecting line. Point value: six points maximum.

4. Connections made with a line (Cl)—Any drawing connection made with a line between one figure fragment or figure and another. One point for each connecting line. Point value: six points maximum.

5. Connections made to produce a theme (Cth)—Any figure contributing to a compositional theme. One point for each theme-bound figure. Point value: six points maximum.

6. Boundary breaking that is fragment dependent (Bfd)—Any extension or continuation of the "small open square" located outside the square frame. Point value: six points.

7. Boundary breaking that is fragment independent (Bfi)—Any drawing independent of the "small open square" that breaks the square frame or is located outside of it. Point value: six points.

8. Perspective (Pe)—Any breaking away from two-dimensionality. One point for each figure; six points for a perspective solution contributing to a compositional theme with perspective. Point value: six points maximum.

9. Humor (Hu)—Any drawing that elicits a humorous response. Point value: six points maximum.

10. Unconventionality (Uc)—Any manipulation of material (e.g., the turning of the testing paper [3 points]). Any surrealistic or abstract elements (e.g., the use of an abstract theme [3 points]). Any combination of figures, signs, or symbols (e.g., the semicircle does not become a stereotypical sun or face [3 points]). Point value: 12 points maximum.

11. Speed (Sp)—A break-down of points according to the time spent on each drawing production:

Under 2 minutes: 6 points
Under 4 minutes: 5 points
Under 6 minutes: 4 points
Under 8 minutes: 3 points
Under 10 minutes: 2 points
Under 12 minutes: 1 point

Scores are entered on a grid such as this:

Scoring Grid

Cm					
Ad					
Ne					
Cl					
Cth					
Bfd					
Bfi					
Pe					
Hu					
Uc	_____ +	_____ +	_____ +	_____ =	
Sp					
DP Total					/72

Activities to Foster Creativity 📖

Creativity is not easy to assess, but creativity can be fostered in a classroom environment in which students are engaged in activities that include projects and scenarios to which they respond. Examples such as the following Floodville sample can be linked with natural disasters, environmental issues, ecosystems, water recreation, water pollution, city planning, economics, mapping skills, governmental involvement, and numerous real-world events.

A sample like Floodville might begin with a scenario created by the teacher or students and teacher. To help in creating this scenario, the newspapers or other media news could serve as a source for details of real-world events, and a scenario that has local relevance can be generated. The local weather history of an area could also be brought into the scenario. Scenario development can be accomplished through four major steps. The teacher may want to model this process with a teacher-generated scenario and drawing, and then students, working in cooperative groups, could also generated their own scenarios.

Floodville (Creativity Domain) 📖

1. Generate a scenario: Floodville is a town that was established on the banks of the Floody River, which is deep enough to allow barge travel and support recreational activities, such as boating and fishing. In fact, many people own homes on river-front property. Floodville sits in a valley surrounded by hills that were once covered with trees. The trees were a source of valuable timber, and the hills are low enough for farming. Floody River floods in the spring but usually causes little or no major property damage and only a few road closings. About every 10 years, a major flood occurs and causes several million dollars worth of property damage.

2. Brainstorm some questions or issues for student reactions: The government proposes to have Floodville relocate to a higher elevation. This would involve moving 250 families, a school, two churches, two small factories, a grocery store, the city offices, and two gas stations. Imagine that you are a Floodville resident: What are some kinds of information would you want to have before you would support such a decision? What would you propose as alternatives to relocation? What rebuttals could be made to support not relocating the town? What changes to the surrounding environment likely worsened the flooding? How might runoff from farms create problems? Would a dam or flood wall help?

FIGURE 5.2

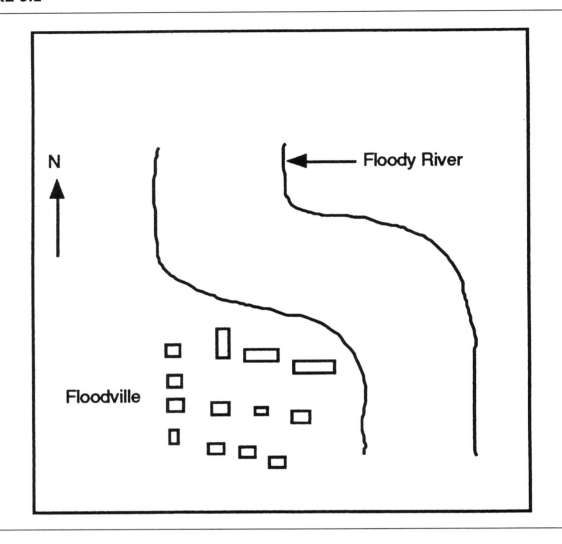

3. Relate to pertinent issues: This kind of scenario can also be moved to issues such as whether building should be permitted on flood plains (or coastal areas subject to hurricanes or in earthquake-prone areas). Role-playing is also a viable option for addressing these issues.

4. Foster creativity: What constitutes creativity in solutions to scenarios would depend on novelty of solutions presented, and this novelty should likely be grounded in the reality of a workable solution. A suggestion would be to focus on assessing problem solving and higher-order thinking skills. Marzano, Pickering, and McTighe (1993), in their dimensions of learning model, provide a set of rubrics on which to assess these skills.

Assessing Attitudes, Preferences, and Processes

The five instruments that follow (Forms 5.5-5.9), taken from various sources, are designed to assess aspects of students not readily observed or accessible through curriculum content.

NOTE

1. This section is based on the work of Jellen and Urban (1986). The Test for Creative Thinking—Drawing Production (TCT—DP) is available from Swets Test Publishers, Lisse, Netherlands; www.swetstest.nl

FORM 5.5

Science, Technology, and Society Attitude Scale (Attitude Domain)

Circle the letter that best matches your view. Use the following rating scale:

a = Strongly Agree, b = Agree Moderately, c = Neutral, d = Disagree Moderately, e = Strongly Disagree

	Strongly Agree	Agree Moderately	Neutral	Disagree Moderately	Strongly Disagree
1. Problems resulting from science or technology hardly ever affect me.	a	b	c	d	e
2. Most people will not act on STS issues even if they understand why action is necessary.	a	b	c	d	e
3. All science classes should incorporate STS issues and topics into the present curriculum.	a	b	c	d	e
4. I would be willing to pay 10% more per product if manufactures would use this money to reduce their pollution.	a	b	c	d	e
5. Science content is the most important aspect of STS education.	a	b	c	d	e
6. I know all I care to know about environmental problems.	a	b	c	d	e
7. STS education should incorporate a problem-solving approach.	a	b	c	d	e
8. The government is really trying to solve our STS problems.	a	b	c	d	e
9. Open discussion should be a major component of STS curriculum.	a	b	c	d	e
10. In the future, there will not be any problems resulting from science or technology in the United States.	a	b	c	d	e
11. Science process skills should be used whenever possible when teaching STS topics.	a	b	c	d	e
12. I want to know how STS issues affect me.	a	b	c	d	e
13. Members within a society have the responsibility to develop an appreciation of, and respect for, the rights of others within the society.	a	b	c	d	e
14. I would like more information on pollution.	a	b	c	d	e
15. If I knew more about STS issues, I would do more about them.	a	b	c	d	e
16. Declining environmental quality poses a serious threat to the future health and well-being of most Americans.	a	b	c	d	e
17. Technology eventually will solve all our problems.	a	b	c	d	e
18. Consumers like myself are really trying to solve the problems dealing with STS.	a	b	c	d	e

SOURCE: Adapted from NAEP (1978).

Enger, S., & Yager, R. *Assessing Student Understanding in Science*. © 2001 by Corwin Press, Inc.

FORM 5.6

Assessing Attitudes and Preferences in Science—Grades 4 Through 12
(Attitude Domain)

For statements 1 through 18, circle the letter that best indicates your view. Use the following rating scale:
a = Always, b = Frequently, c = Sometimes, d = Rarely, e = Never

	Always	*Frequently*	*Sometimes*	*Rarely*	*Never*
1. Science classes are fun.	a	b	c	d	e
2. Science classes increase my curiosity.	a	b	c	d	e
3. The things studied in science classes are useful to me in daily living.	a	b	c	d	e
4. Science classes help me test ideas I have.	a	b	c	d	e
5. My science teacher frequently admits to not having answers to my questions.	a	b	c	d	e
6. Science classes provide me with skills to use outside of school.	a	b	c	d	e
7. My science class deals with the information produced by scientists.	a	b	c	d	e
8. Science classes are exciting.	a	b	c	d	e
9. Science classes provide a chance for me to follow up on questions I have.	a	b	c	d	e
10. Science teachers encourage me to question.	a	b	c	d	e
11. All people can do and practice basic science.	a	b	c	d	e
12. Being a scientist would be fun.	a	b	c	d	e
13. Being a scientist would make a person feel important.	a	b	c	d	e
14. Science classes are boring.	a	b	c	d	e
15. Being a scientist would be lonely.	a	b	c	d	e
16. Being a scientist would make a person rich.	a	b	c	d	e
17. Being a scientist would mean giving up some things of interest.	a	b	c	d	e
18. Scientists discover information that is difficult to understand.	a	b	c	d	e

For items 19 through 26, circle "a" for the subjects you like best, and
circle "b" for the subjects you do not like as well. Last, circle the name of your favorite subject.

19. Foreign Language	a	b		23. Physical Education	a	b
20. Science	a	b		24. Language Arts	a	b
21. Mathematics	a	b		25. Reading	a	b
22. Social Studies	a	b		26. Music	a	b

SOURCE: Adapted from NAEP (1978).

Enger, S., & Yager, R. *Assessing Student Understanding in Science.* © 2001 by Corwin Press, Inc.

FORM 5.7

Science Process Test

Name _____   Date _____

Teacher _____   Course _____

Circle the letter that best applies to you.

1. Gender: a = female   b = male   a   b

2. Ethnicity: a = Asian   b = Black   c = Hispanic, Latin   a   b   c   d   e
   d = Caucasian   e = Native American, Alaskan Native
   Other (please write in) _____

3. Grade: a = 6, 7, 8   b = 9   c = 10   a   b   c   d   e
   d = 11   e = 12

For each question, circle the letter of the best answer.

Making Observations

4. Which of the following is an observation only?
   A. The piece of metal is red so it must be hot.
   B. The street is wet so it must have rained.
   C. The table looks like it is made of wood.
   D. The block is orange.

5. Which of the following could be observed with the sense of sight?
   A. A change in temperature of the air.
   B. A change in height of plants.
   C. A change in sweetness of a new chemical.
   D. A change in the noise made by an engine.

Using Space-Time Relationships

6. If runners A and B start at the same time and they arrive at the finish line (C) at the same moment, who ran faster?
   A. A ran faster than B.
   B. B ran faster than A.
   C. A and B ran at the same speed.
   D. B ran slower than A.

FORM 5.7 Continued

7. Which shape of shadow could *not* be formed by a solid cylinder?
   A. A circle
   B. A square
   C. A rectangle
   D. A triangle

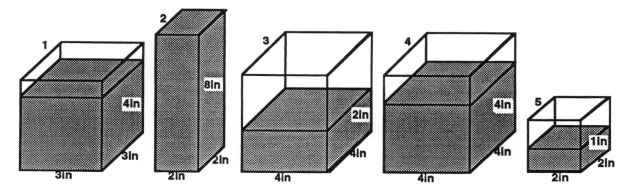

8. According to the preceding diagram, which two containers have approximately equal volumes of water in them? (Drawings are not to scale.)
   A. Containers 1 and 2
   B. Containers 2 and 3
   C. Containers 3 and 5
   D. Containers 2 and 5

Classification

9. The following table has some background information about students in Jones Elementary School.

1. Name	2. Gender	3. Birthday	4. Nationality	5. Year Entered School
M. John	Female	June 1990	American	1996
B. Thames	Male	March 1990	British	1996
A. Shirley	Male	December 1989	American	1996
R. Thompson	Female	May 1990	American	1996
R. Ali	Male	October 1989	Indonesian	1996
B. Ghombal	Male	August 1989	Portuguese	1996

   If you wanted to sort these students into at least *two* different groups, which of the following categories could *not* be used to do so?
   A. Gender (male or female)
   B. Year of birth
   C. Nationality
   D. Year entered school

(continued)

Enger, S., & Yager, R. *Assessing Student Understanding in Science.* © 2001 by Corwin Press, Inc.

FORM 5.7 Continued

10. Which would be the best feature to use in classifying the following shapes into two groups?

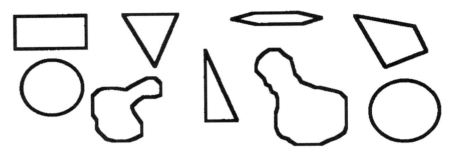

    A. Square versus not square.
    B. No straight sides versus four straight sides.
    C. Circle versus triangle.
    D. Curved edge versus straight edge.
    E. Odd number of sides versus even number of sides.

11. The hotter the water, the faster sugar will dissolve. Look at the jars in the following drawing. Each jar has the same amount of sugar. Put the jars in order from the slowest rate for sugar to dissolve to the fastest rate to dissolve.

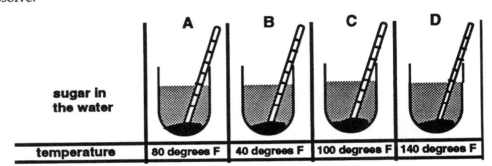

    A. A, B, C, D
    B. B, A, C, D
    C. C, B, D, A
    D. D, C, B, A

Using Numbers

12. Which of the following groups of objects presents the objects in order, from left to right, of the smallest to the largest number?

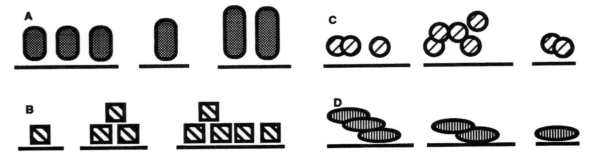

FORM 5.7 Continued

13. What is the next number in this number sequence?

    2   3   5   8   12   17   ?

    A. 19
    B. 23
    C. 24
    D. 28

14. Yesterday it was -5° C. Today it is 10° C. How many degrees warmer is it today than it was yesterday?

    A. 5 C° warmer
    B. 10 C° warmer
    C. 15 C° warmer
    D. -5 C° warmer

Measuring

15. Normal human body temperature is about 37°C. The body temperature of someone who is ill ranges from about 36°C to 42°C. Which thermometer would be the best to use for measuring any human body temperature?

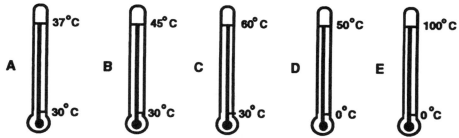

    A. A
    B. B
    C. C
    D. D
    E. E

16. Four students were each given their own plant. To practice their measuring skills, each student was asked to measure the height of the plant four times during a single class period. Which student do you think measured most carefully and precisely?

Student	Measurement Number			
	1	*2*	*3*	*4*
Rusty's plant heights	3 cm	6 cm	10 cm	8 cm
Mike's plant heights	4 cm	5 cm	5 cm	4 cm
Karen's plant heights	2 cm	10 cm	4 cm	8 cm
Carol's plant heights	8 cm	3 cm	2 cm	1 cm

    A. Rusty
    B. Mike
    C. Karen
    D. Carol

(continued)

Enger, S., & Yager, R. *Assessing Student Understanding in Science.* © 2001 by Corwin Press, Inc.

FORM 5.7 Continued

17. Dan and Tom each shot 20 foul shots on each day for days. Using the following graph, on how many days did Dan make more baskets than Tom?

Baskets

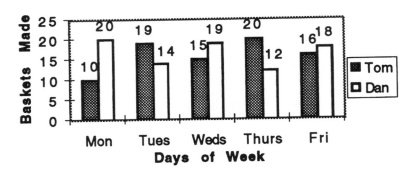

A. 1 day
B. 2 days
C. 3 days
D. 4 days
E. 5 days

Communicating

18. What object has 6 equal faces, 8 corners, 12 edges, and volume?
   A. A cube
   B. A square
   C. A sphere
   D. A cone
   E. A hexagon

19. A tennis ball was dropped from several different heights, and the height the ball bounced was recorded each time the ball was dropped. Which of the following would be the best method to report the data collected?
   A. A written paragraph
   B. A tally of the number of bounces
   C. A frequency distribution
   D. A bar graph
   E. A pie chart

20. Mary wants to make a diagram of the school classroom on a piece of notebook paper. Which of the following choices would be a convenient scale for her to use?
   A. 1 inch = 1 mile.
   B. 1 inch = 1 centimeter.
   C. 1 inch = 1 yard.
   D. 1 inch = 1 acre.
   E. 1 inch = 1 inch.

FORM 5.7 Continued

Making Inferences

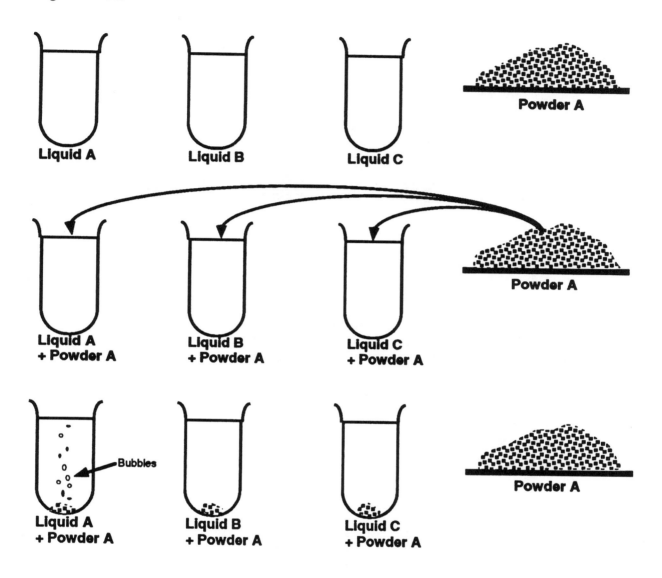

21. Which inference is best supported by the above diagram?
    A. Liquids A and C are the same.
    B. Liquids A and B are not the same.
    C. Liquids B and C are both the same.
    D. Liquids A, B and C are all the same.

(continued)

FORM 5.7 Continued

Predicting

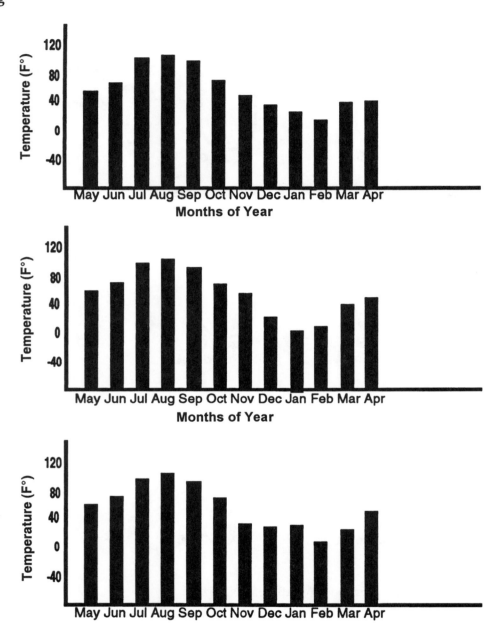

22. The average temperature recorded each month during the past 10 years is shown on these graphs. Based on this evidence, which month do you think will be coldest next year?

   A. June
   B. September
   C. November
   D. January
   E. February

Enger, S., & Yager, R. *Assessing Student Understanding in Science.* © 2001 by Corwin Press, Inc..

FORM 5.7 Continued

23. The same amount of a gas that is less dense than air is placed in each of the following balloons. Which balloon do you think will float up the fastest?

Weight: A = 1000 pounds   B = 800 pounds   C = 500 pounds   D = 200 pounds

24. Look at the following objects below. Which item do you think will sink fastest in a pan of water?

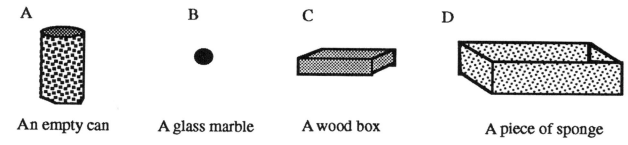

A   B   C   D

An empty can   A glass marble   A wood box   A piece of sponge

   A. An empty can
   B. A glass marble
   C. A wooden box
   D. A piece of sponge

25. Dan and Dawn want to know if there is any difference between the mileage expected from bicycle tires from two different manufacturers. Dan will put one brand on his bike, and Dawn will put the other brand on her bike. Which of the following variables would be *most* important to control in this experiment?
   A. The time of day the test is made
   B. The number of miles traveled by each type of tire
   C. The physical condition of the cyclist
   D. The weather conditions
   E. The weight of the bicycle used

26. A group of students conducted an experiment to determine the effect of heating on the germination (sprouting) of bean seeds. Which of the variables in the following list is *least* important to control in this experiment?
   A. The temperature to which the seeds are heated
   B. The length of time the seeds are heated
   C. The type of soil used
   D. The amount of moisture in the soil
   E. The size of the container used for growing each seed

(continued)

Enger, S., & Yager, R. *Assessing Student Understanding in Science.* © 2001 by Corwin Press, Inc.

FORM 5.7 Continued

27. A student wants to know how the amount of acid rain affects the fish population. She takes two jars and fills each of the jars with the same amount of water. She adds 50 drops of vinegar (acid) to one jar and adds nothing extra to the other. She then puts 10 similar fish in each jar. Both groups of fish are cared for (oxygen, food, etc.) in identical fashion. After observing the behavior of the fish for a week, she makes her conclusions. Without adding another variable, what could you suggest to improve this experiment?
   A. Prepare more jars with different amounts of vinegar (acid).
   B. Add more fish to the two jars already in use.
   C. Add more jars with different kinds of fish and different amounts of vinegar in each jar.
   D. Add more vinegar to the two jars already in use.

Interpreting Data

28. The following data are taken from an experiment:

Temperature (average)	Mass of Seed (grams)	Water Uptake (ml/day)	Exposure to Light (min/day)	Plant Height (cm/20 days)
20°C	2.2	10	20	20.2
50°C	2.3	10	20	20.3
30°C	2.3	10	20	20.2
25°C	2.1	10	20	20.3
25°C	2.3	10	30	21.9
25°C	2.2	10	40	22.8
20°C	2.2	10	30	21.8
20°C	2.1	20	30	21.9
20°C	2.2	30	30	22.0

Based on the data in the table, what factor do you think has the greatest influence on the rate of plant growth?
   A. The temperature where the plant is grown
   B. The seed mass
   C. The amount of water uptake each day
   D. The length of the time the plant is exposed to the light

29. Here is an experiment that shows how much some peanut plants grew in 20 days.

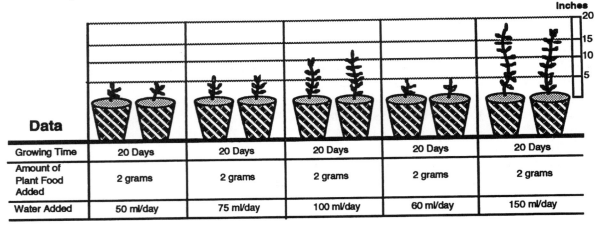

Data					
Growing Time	20 Days	20 Days	20 Days	20 Days	20 Days
Amount of Plant Food Added	2 grams	2 grams	2 grams	2 grams	2 grams
Water Added	50 ml/day	75 ml/day	100 ml/day	60 ml/day	150 ml/day

FORM 5.7 Continued

Look at the information given in the graphic. Which is the best conclusion for this experiment?
A. The more plant food that was added, the faster the plant grew
B. The more plant food that was added, the slower the plant grew
C. The more water that was added, the faster the plant grew
D. The more water that was added, the slower the plant grew

Formulating a Hypothesis

30. Bob set up two identical bowls. Both contained sugar water, and both were open to the air. One was placed in the dark, whereas the other was placed in the light. What one item differs from one set-up to the other?
A. The exposure to light
B. The shape of the bowl
C. The exposure to air
D. The amount of sugar

31. Which of these statements is the best example of a hypothesis?
A. This magnet picked up 12 paper clips.
B. The milk in this bottle froze in 20 minutes.
C. The house plant may have died from being watered too much.
D. The leaves on that maple tree have all turned red.
E. At that rate, the pool filled in 10 minutes.

32. Examine the following data table, and select the most appropriate hypothesis regarding dissolving time and water temperature.

	Average dissolving time (in seconds)			
Substance	Water 20°C	Water 40°C	Water 50°C	Water 60°C
20 g of sugar	80	40	20	5
20 g of salt	60	30	16	3

A. There is no difference in dissolving time of the substances because of water temperature.
B. The lower the temperature of water, the less time needed to dissolve the substances.
C. The higher the temperature of the water, the less time needed to dissolve the substances.
D. It is impossible to make a hypothesis from the information given in the chart.

Defining Operationally

33. Which one of the following *is* written as an operational definition?
A. Because the density of oil is less than the density of water, when water is mixed with oil, the oil will float on the surface of the water.
B. The speed of a supersonic jet is similar to the speed of sound waves.
C. When you drive your car at a speed of 30 miles per hour, you have to push the brake pedal 300 feet before the line or point you are planning to stop.
D. The speed of a car will decrease when it has to turn right or left.

(continued)

Enger, S., & Yager, R. *Assessing Student Understanding in Science.* © 2001 by Corwin Press, Inc.

FORM 5.7 Continued

Experimenting

34. A student wants to test to see if the color of cloth influences the amount of heat absorbed. He plans an experiment using two colors of cloth to wrap two different glasses, each containing the same amount of water. One glass is wrapped with green cloth, and the other is wrapped with yellow. He places the glasses in a sunny spot and sets a thermometer in each glass to observe the temperature. What things can you suggest to improve his testing?
   A. Add to the number of glasses covered with the cloth
   B. Reduce the amount of water in each glass
   C. Prepare more containers each covered with a different color of cloth
   D. Double the size of the cloth used to cover the glass

35. Eight bean seeds were allowed to germinate and then divided into four groups of two seeds each. One group was grown under red light, another under green light, another under blue light, and the fourth under white light. At the end of 2 weeks, the growth of each group of plants was measured to see which group of plants had grown the most. This experiment could *best* be improved by doing which of the following?
   A. Giving more water to the plants grown under the red light
   B. Increasing the number of seeds grown in each of the four groups
   C. Growing just the plants under white light in sandy soil, but growing all others in humus soil
   D. Adding one more group of two seeds to the experiment and growing them under purple light

36. Gloria wants to determine the temperature best suited for fish. Which of the following procedures is the best to use to determine this?
   A. Get 5 aquaria (fish tanks) and place 5 similar fish in each aquarium. Keep the temperature in each aquarium constant at 25°C.
   B. Place 5 fish in one aquarium. At intervals of 10 minutes, change the water temperature from 10°C to 15°C; to 20°C; to 25C; and finally to 30°C. Observe the behavior of the fish after each change in temperature.
   C. Get 5 aquaria and place 5 similar fish in each aquarium. Keep the temperature of the water constant at about 25°C, and observe the behavior of the fish in each aquarium.
   D. Get 5 aquaria, and place 5 similar fish in each aquarium with the temperature of the water varied from 15°, 20°, 25°, 30°, to 35°C in each aquarium. Observe the behavior of the fish in each aquarium.

SOURCE: This form is adapted from the Mason City Schools, Mason City, Iowa, 1993, and from NAEP (1978).

Enger, S., & Yager, R. *Assessing Student Understanding in Science.* © 2001 by Corwin Press, Inc.

FORM 5.8

Public Release of 1990 NAEP Assessment Items for Nature of Science
(Concept and Nature of Science Domains)

Questions 17 through 22 are appropriate for Grades 9 through 12.

For each question, circle the letter of the best response.

1. Knowledge of earth's past continues to change as scientists find additional fossils. This is because
   A. Scientific knowledge cannot be trusted.
   B. Scientists change their ideas as new evidence is found.
   C. Scientists do not accurately report what they observe.
   D. Fossil study is not a true science.

2. Which of the following statements about scientific knowledge is correct?
   A. It is based on observations and experiments that can be repeated by scientists.
   B. It cannot be tested.
   C. It is based on laws that never change.
   D. It is based on beliefs and faith.

3. Hypotheses are
   A. Ideas that can be tested.
   B. Facts about science.
   C. Observations of nature.
   D. Results of experiments.

4. Scientists estimate that Earth is about how many years old?
   A. 100 billion years old
   B. 5 billion years old
   C. 60 million years old
   D. 1 million years old
   E. 10 thousand years old

5. The methods of science can be used to answer all of the following questions *except*
   A. How many butterflies are there in California?
   B. What are some of the effects of rainfall on the growth of roses in Kentucky?
   C. Are roses more beautiful than butterflies?
   D. Which brand of paper towels absorbs the most water?

6. Which of the following questions would be the easiest to answer with an experiment?
   A. How many uses are there for magnets?
   B. Which is the stronger of two magnets?
   C. What makes a magnet strong?
   D. How are magnets made?

(continued)

FORM 5.8 Continued

Use the following graph to answer question 7.

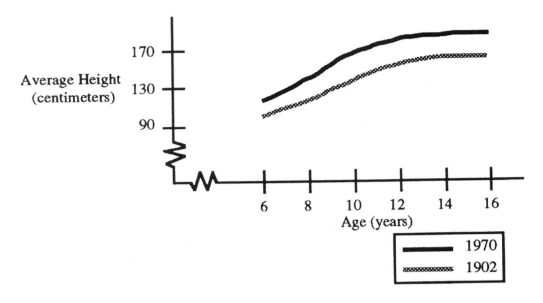

7. Which statement is best supported by the graph?
   A. On average, children in 1970 grew taller than children in 1902.
   B. On average, children in 1902 grew taller than children in 1970.
   C. There is no difference between 1902 and 1970 in the average height of children.
   D. Younger children are taller than older children.

8. If José wants to find out how fast a pumpkin vine grows, which of the following should he do?
   A. Sprout some pumpkin seeds in a cup.
   B. Measure the longest pumpkin vine he can find.
   C. Measure length of a pumpkin vine each week while it is growing.
   D. Give one pumpkin vine more water and fertilizer than another pumpkin vine and compare the results.
   E. Measure the diameter of the largest pumpkin he can find.

9. You notice a flat, open container lying on the beach. There is a small amount of water in the center of the container and a layer of crusty sediment around the water. You think the crusty material probably came from evaporation of the water. This is an example of
   A. An observation
   B. A hypothesis
   C. A variable
   D. An experiment

10. Which of the following is an opinion rather than an observation?
   A. Many plants are green.
   B. Many flowers are beautiful.
   C. Plants require sunlight.
   D. Plants can grow in different places.

FORM 5.8 Continued

The diagram to the right represents the fossil evidence of an event that took place far back in the Earth's history.

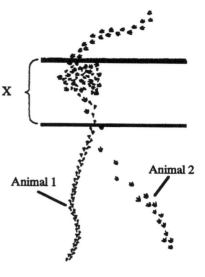

11. Which of the following probably took place in area X of the diagram?
    A. Animal 2 was eaten by Animal 1.
    B. Animal 2 flew away.
    C. Animal 1 was eaten by Animal 2.
    D. Animal 1 stepped in the tracks of Animal 2 when leaving area X.

12. A group of students is told to put 30 fossils into several groups. The students find it difficult to agree on one solution. The main reason that scientists often have the same problem is that
    A. Fossils have not been studied very much.
    B. They work in groups.
    C. Fossils are very old.
    D. There are many different ways to classify fossils.

13. Several students notice that the river running through a park is shallower than it was 1 week before. One student says that this change must be due to a lack of rain. This statement is best described as
    A. A hypothesis that can be investigated.
    B. An observation based on theory.
    C. A conclusion based on firm evidence.
    D. A theory based on experimentation.

14. A student observed a spider and its web. Which of the following is *not* an observation?
    A. The web has some threads that are straight.
    B. The spider has eight legs.
    C. The spider's abdomen is larger than its head.
    D. The spider makes no noise.
    E. The spider evolved from insects.

15. Jamal has 20 silk worm larvae. Half are 2 centimeters long and half are 4 centimeters long. He knows the length of time it takes the smaller larvae to consume 100 grams of mulberry leaves. Which of the following kinds of information should he collect for the 4-centimeter larvae in order to compare the eating rates of the two sets of larvae?
    A. Time for all 10 larvae to eat 100 grams of leaves
    B. Time for each larva to eat one leaf
    C. Weight of leaves eaten by all 10 larvae in an hour
    D. Number of leaves eaten by all 10 larvae in a day

16. A computer company wanted to know how often American eighth-grade students used a computer in earth science class. What is the most efficient way to get this information?
    A. Ask all eighth graders in the country
    B. Ask some eighth graders in one school district
    C. Ask all eighth graders in one state only
    D. Ask some eighth graders in 25 states

(continued)

Enger, S., & Yager, R. *Assessing Student Understanding in Science.* © 2001 by Corwin Press, Inc.

FORM 5.8 Continued

17. Measurements taken during a scientific experiment should be both accurate and precise. Accuracy refers to the
    A. Closeness of the measurements to the true value
    B. Reproducibility of the measurements
    C. Location of the measurements taken
    D. Time between measurements taken
    E. Number of measurements taken

18. A scientist develops a theory to explain some phenomena that previous theories could not. However, this theory leads to predictions that are contrary to other scientists' expectations. What should be done in response to these results?
    A. Ignore the expectations and accept the theory.
    B. Reject the theory because it is contrary to the expectations.
    C. Revise the theory so that it agrees with the expectations.
    D. Design experiments to test for the predictions made by the theory.
    E. Develop another theory that predicts what the scientists expected.

19. Which of the following statements can be tested most easily?
    A. The Earth's temperature in 2004 will be colder than in 2003.
    B. Physics is more interesting than biology.
    C. DNA is more important than protein.
    D. Monkeys are more insightful than cattle.
    E. Old seeds are less likely to germinate than new seeds.

20. While performing an experiment, a scientist finds evidence that a new element exists. Which of the following should the scientist do first?
    A. Publish the finding to be first.
    B. Repeat the experiment using the same conditions.
    C. Repeat the experiment but change one variable to see whether that affects the results.
    D. Wait and see whether another scientist finds the same evidence.
    E. Look for additional new elements that are similar to the new one.

21. Which of the following procedures is essential in an experimental study designed to investigate the effects of vitamin K in the diets of humans?
    A. Make sure that all the study subjects get the same amount of vitamin K.
    B. Use only students as subjects.
    C. Use several different brands of vitamin K.
    D. Make sure that all study subjects are kept in different environments.
    E. Divide the study subjects into experimental and control groups.

22. Currently, scientists make many predictions about the way solids behave by assuming that the atoms in a solid behave as if they are connected by tiny springs. This means that which of the following is true?
    A. There must be tiny springs connecting atoms together.
    B. Scientists will always find it useful to think of atoms as connected by springs.
    C. An intelligent life form on another planet would almost certainly think about solids in the same fashion.
    D. Thinking of atoms as being connected by springs constitutes a useful model of solids.
    E. Because it is a successful model, the spring model of the atom can be classified as a fact.

SOURCE: Adapted from the Mason City Schools, Mason City, Iowa, 1993.

FORM 5.9

NAEP Questionnaire for Student Views about Scientific Theories and Scientists
(Nature of Science Domain)

For each question, circle the letter(s) which best indicate(s) how you feel about the statement.

SA = Strongly Agree, A = Agree, N = Neither Agree nor Disagree, D = Disagree, SD = Strongly Disagree

A: How do you feel about each of these statements about scientific theories?

1. It is likely that some theories that scientists use today will be shown to be inadequate someday.   SA   A   N   D   SD

2. Scientists are interested in improving their explanations of natural events.   SA   A   N   D   SD

3. Theories are useful even though they may be incomplete.   SA   A   N   D   SD

4. Scientific theories are important products of science.   SA   A   N   D   SD

5. One important use of a scientific theory is to predict future events.   SA   A   N   D   SD

B: How do you feel about each of these statements about scientists?

6. One very important job of a scientist is to report exactly what he or she observes.   SA   A   N   D   SD

7. Scientists should not criticize each other's work.   SA   A   N   D   SD

8. If a researcher accurately reports his or her experimental results, other researchers should accept the results without question.   SA   A   N   D   SD

9. Once scientists have developed a good theory, they should stick together to prevent others from saying it is wrong.   SA   A   N   D   SD

10. Scientists must be willing to change their ideas when new information becomes known.   SA   A   N   D   SD

11. Different scientists may give different explanations about the same observations.   SA   A   N   D   SD

SOURCE: Adapted from the Mason City Schools, Mason City, Iowa, 1993.

6

Assessment Samples for Grades K Through 4

In this chapter, you can find suggestions for

- Structuring a fair test

- Asking concept-related application questions

- Assessing student attitudes toward science

- Student self-assessments

- Assessing students' perceptions of scientists

Applying Process Skills and Experimental Design

📖 *Does the Brand of Bubble Gum Make a Difference in Bubble Size?*

The use of process skills can be supported by framing experiences in a context to which children can relate. For example, if students wanted to know which brand of bubble gum could be used to blow the biggest bubbles, they might first ask the obvious question, "Which brand of bubble gum makes the biggest bubbles?" The second question asked might be, "How could we study this question?" From here, the students and the teacher could set up a plan of action to try to find out.

In setting up a plan of action to address the question of interest, class discussion could be guided toward conducting a fair test. Students might brainstorm a list of things that would need to be considered, such as

- How many brands of gum should we test?

- Does sugar make a difference?

- How long should we chew the gum before blowing a bubble?

- What styles or techniques of bubble blowing should we use?

With younger students, the idea of controlling variables may be too abstract, but a discussion of why they would want to try to keep things the same when doing a fair test would seem appropriate. The students could design a plan together in which they would all use the same brand of sugarless gum. They could decide on how long to chew the gum and then try a technique to blow a bubble. Students could be asked to estimate how big a bubble they thought they could blow. Students could be asked to really focus on the technique that they used and then write descriptions of their techniques.

Students could then go on to test two brands of sugarless gum. This could be done as a cooperative group project in which one student could be the bubble blower, one the data collector, one the recorder, and one to be the quality control agent to see that the plan is followed. When students have completed collecting their data, they could be asked to average their data and write about their findings. All of the class data could be displayed, discussed, and summarized. Students could also write about their findings and develop a report to send to the bubble gum manufacturer.

A checklist approach (see the following table) could be used to document class progress in process skill use and designing experiments.

			Improvements Needed
Yes	No	Students were able to generate a question to investigate.	
Yes	No	Students were able to generate a plan to follow.	
Yes	No	Students were able to set up a fair test.	
Yes	No	Students were able to set up a data table.	
Yes	No	Students were able to conduct a fair test.	
Yes	No	Students were able to record data.	
Yes	No	Students used data-gathering tools in appropriate ways.	
Yes	No	Students were able to tabulate their data.	
Yes	No	Students were able to graph their data.	
Yes	No	Students were able develop a conclusion and report their findings.	

Application Assessment Items for Grades K Through 3

Form 6.1 presents some sample assessment items for the application domain, appropriate for Grades K through 3.

Assessing Attitudes About Science

Forms 6.2 and 6.3 present ways to elicit young students' attitudes about science. Form 6.2 uses visual cues, and Form 6.3 is to be used in conjunction with interviewing.

Student Self-Assessment

After completing an activity on measuring various objects, students can indicate what they were able to do on an instrument such as the one shown in Form 6.4; they should have their work available so they can share what they have done. Students could be asked to draw a star by the things that they can do. This is a way to engage students in self-assessment at an early age. The NSES (NRC, 1996) recommends that students use conventional measuring instruments, such as rulers, thermometers, and balances, and the student checklist could be modified according to the instruments used. In science, measurements are usually metric based, but with younger students, the use of the English system may be appropriate as their initial frame of reference. The desired units could indicate those related to the activity.

Assessing Perceptions About Scientists

Students may have stereotypic views of scientists and what scientists do. The following assessment could be used at the beginning of the year and again at the end of the year to see if student perceptions have changed. For students' perceptions to change, it is likely that classroom instruction would need to address misperceptions about scientists. By having students draw pictures of their ideas of a scientist, a look at students' perceptions is possible.

An "if-then" to consider: If children hold stereotypic views of scientists and what scientists do, then what will you, as the teacher, do to change these perceptions? Fort and Varney (1989) reported on a study in

(text continues on page 135)

FORM 6.1

Sample Application Assessment Items: K–3
(Application Domain)

1. Concept: Birds hatch from eggs.

   Application Question: You have found some eggs in a nest in a tree. You watch the nest and the eggs hatch. Draw a picture of the kind of animal that you think will come from the eggs. (A picture could be used to elicit student responses.)

2. Concept: The cyclic movement of water between the land and the air is called the water cycle.

   Application Question: Yesterday, you made a mud pie while playing outside. The pie was very runny, so you left it outside. If tomorrow is sunny, draw a picture of what it will look like when you go out to play with it.

3. Concept: When resistance is increased, it takes more force to move an object.

   Application Question: Which child is working harder to move the object?

   (Two similar pictures are needed. One should be of a child pulling an empty sled or wagon and the second picture would show an object placed on the sled or in the wagon.)

4. Concept: Environments and habitats are all changing. Some changes are a result of natural forces, and some are a result of the actions of people.

   Application Question: A company is going to build a road on some land with forest. Draw one picture of what you think the land looked like before the road was built. Then draw a picture of how the area would look when the road was being built.

5. Concept: Meters are used for measuring distance.

   Application Question: The teacher has asked you to measure the length of your classroom. What unit of measure will you use? Circle the best answer.

   Kilometers   Meters   Grams   Liters

Enger, S., & Yager, R. *Assessing Student Understanding in Science.* © 2001 by Corwin Press, Inc.

FORM 6.2

How I Feel About Science: Grades K–3

Name: _____ Grade Level: _____

How do you feel about your science class? After each sentence, place an X on the face that shows how you feel.

	Yes	It's OK	No
1. Science time is fun.	☺	😐	☹
2. Things I learn in science help me understand things at home.	☺	😐	☹
3. I can do science things now.	☺	😐	☹
4. We do fun things in science class.	☺	😐	☹
5. In science, my teacher likes to have me ask questions.	☺	😐	☹
6. I might want to do a science job when I grow up.	☺	😐	☹

FORM 6.3

How I Feel About Science: Grades 2-4

What are some of your thoughts about science?

1. What would you write or say to finish this sentence?

   To me science is . . .

2. Tell about all the ways you use science in a day.

3. What are some things you like about science?

4. What are some things you do not like about science?

5. If someone told you that science is not important, tell why you would agree or disagree with them.

6. If you woke up tomorrow morning and were in the Land of Science, describe what you think it would be like, how you would feel, and what you would do.

FORM 6.4

I Can Measure With A Ruler! (Integrates Domains)

1. I can start from the left end of the ruler.

2. I can measure in inches.

3. I can measure in feet.

4. I can measure four (4) things.

5. I can draw a picture of each thing I measured.

6. I can write the measurement under each picture.

which students were asked to portray scientists in words and pictures and found that based on the data, students mostly viewed scientists as male, white, and benevolent. Barnum (1996, 1997) invited science teachers to participate in a study in which their students drew scientists and then wrote about them. Barnum also suggested asking students to draw themselves doing science and recommended conducting taped interviews with students.

📖 *Draw a Scientist (Nature of Science Domain)*

Purpose: To gain insight into how children view scientists and what they think scientists do.

Preassessment: Ask students to draw a picture of a scientist. Also ask them to describe what they have drawn and tell what they think a scientist does.

Interim Instruction: If students hold stereotypic views of scientists, then science instruction must include experiences that foster changes in student thinking about scientists. Teachers could work to do this by recognizing opportunities that provide evidence that scientists are people and that doing science is a human endeavor. Ideally, raising student awareness of the diversity that exists among scientists would be an ongoing effort in the classroom.

Postassessment: Again ask students to draw a picture of a scientist. Also ask them to describe what they have drawn and tell what they think a scientist does. This postassessment could be used at any time, but it may be most useful at midyear or at the end of the year.

Evaluation of the drawings: Students could be asked to describe any differences between the preassessment and postassessment pictures. If their perceptions have changed, they could also be asked to give reasons for the changes. Barnum (1996, 1997) and Fort and Varney (1989) offer guidelines for evaluating students' drawings.

7

Assessment Examples for
Grades 5 Through 8

In this chapter, you can find suggestions for

- Structuring performance tasks

- Using a context to anchor multiple-choice items

- Incorporating electronic or video portfolios

- Involving parents in portfolio review

- Structuring peer assessment for cooperative groups

- Structuring group assessment of cooperative roles

- Structuring class assessment of group project presentations

- Using student self-assessment of role in group project

Developing a Group Performance Task

📖 *Evolution of Toys: Back to the Future* *(Integrates Domains)*

Developing and crafting a performance task is in itself an evolutionary process. This performance task core idea could be extended in numerous ways to incorporate a variety of skills and disciplines, and as a rule, good instructional tasks have the potential to become good performance tasks.

Performance Task Core Idea

A major toy company has selected the theme "Evolution of Toys: Back to the Future." This toy company has sent out a request for toy designs, and your team of four classmates is interested in designing and submitting a toy. Before your team actually designs a toy, you want to know what toys were like 60 years ago. Using your experiences with toys and other resources, you first plan to compare the toys of today with toys of 60 years ago.

Based on the technology used in the toys children played with in the 1940s, describe three ways that the life of a child of the 1940s and your life in the 2000s would vary. Your team should prepare a two- to three-page written paper that describes two different toys on which you base your predictions about the life of a child in the 1940s. Include a drawing or picture of each of these toys and tell how the toy worked. Your group will also be expected to give a 5-minute oral presentation about your findings to the rest of the class.

Performance Expectations

The following would be possible performance expectations:

- Based on the technology used in the toys, predictions are made about how a child's life in 1940s would differ from a child's life today.

- Predictions are supported by a variety of references and resources.

- References and resources are documented in a bibliography.

- A two- to three-page word-processed paper is developed.

- Drawings of toys and descriptions of how the toys work are included.

- A 5-minute oral presentation is to be made by the group.

- Each member of the group is expected to contribute to the presentation.

- In group work, intrapersonal skills could be evaluated.

- Unique approaches and creativity in products could be noted.

Possible Extensions of the Core Task

- The toys could be drawn to scale.

- Scale drawings could be completed using a computer platform.

- Models of toys could be constructed.

- Production costs for building the toys could be calculated.

- Marketing brochures could be designed.

- The toys could be used as the context for physics concepts.

- The history of the 1940s could be addressed.

- Creative elements could be examined

- Students could design a new toy.

📖 *An Individual or Group Performance Task: Taking Care of Garbage (Integrates Domains)*

✐ *Performance Task Core Idea:*

You have watched the garbage cans being picked up in your neighborhood. Knowing hundreds of garbage cans must be lifted each day, you intend to make life easier for the workers by designing a mechanical device that will lift the garbage can and empty its contents into the garbage truck. Your task is to design and make a working cardboard model of your device and then demonstrate and explain its operation as a machine.

✐ *Performance Expectations*

The following would be possible performance expectations:

- A scale drawing should be completed.

- A model should be constructed.

- A demonstration of the model is given.

- The physics and engineering concepts of the model could be explained.

- Creative elements could be examined.

✐ *Possible Extensions for the Core Task*

- References and resources are documented in a bibliography.

- Connections to the history of technological design and the Industrial Revolution could be made.

- A marketing brochure could be developed.

- The model could be evaluated by an engineer.

- An investigation of the patent process could be completed.

- Famous inventors could be incorporated.

- The creative process could be addressed.

Preparation for Standardized Tests

Preparing students for standardized tests can include practice with a context followed by a selection of multiple-choice questions. This type of format—a contextual setting accompanied by multiple-choice questions—tends to support questions with greater cognitive demands. The following example, adapted from the GED (1990), presents background information along with a table containing the essential facts and pertinent multiple-choice questions. An asterisk marks the best answer for each question.

Multiple-Choice Application Example: 📖 *Concepts of ABO Blood Types (Application Domain)*

Contextual Background for ABO Blood Types ✎

Blood transfusions were practiced in the 1800s, and blood was transferred from one person to another. Sometimes, these transfusions were miraculously successful, whereas at other times, persons experienced pain, tingling sensations, and often, death. In 1901, the scientific basis for successful blood transfusions was identified, and it was recognized that only certain blood types could be successfully transfused to a given individual.

If a person is given a transfusion of an incompatible blood type, his or her blood cells will agglutinate or clump together. Agglutination is due to a reaction between antigens and antibodies in the blood. An *antigen* refers to any substance that produces an immune response, and usually, the body reacts by producing antibodies. The antibodies attack the antigens, which are recognized as foreign substances.

The ABO system is one system used to categorize blood cell antigens. The surfaces of red blood cells of types A, B, and AB contain certain substances that act like antigens. The safest transfusion of blood takes place when the donor and the recipient have the same type of blood. The most critical factors are the antigens present in the donor's blood and the antibodies present in the recipient's blood plasma. If the red blood cells containing antigens are transferred into a person whose blood plasma contains antibodies against them, clumping of the blood occurs.

Use the following information to answer Questions 1 through 5:

Blood Type	Red Blood Cells Have	Plasma Carries
A	A antigen	Antibodies against B
B	B antigen	Antibodies against A
AB	Both A and B antigen	No antibodies against A or B
O	No A or B antigen	Antibodies against A and B

1. A blood bank had five problems with a laboratory technologist's work performance during the first year. Of the problems listed below, which would the technician's supervisor likely consider the most serious?
    a. Coming to work late
    b. Spilling five units of blood
    c. Working too slowly
    *d. Mislabeling four units of blood
    e. Breaking the lens on a microscope

2. From the information presented, the most probable reason some people died in the 1800s after receiving blood transfusions, whereas others recovered, is due to
    a. The type of illness the person had
    b. The skill of the physician
    c. How sterile the equipment was
    *d. Whether or not the person received compatible blood
    e. How much blood was lost during the operation

Suppose that each of the persons below received a blood transfusion:
    A. A four-year-old child
    B. A middle-aged man
    C. An elderly woman

3. In which of these people would agglutination probably occur if given the wrong blood type?
    a. A only
    b. B only
    c. C only
    d. B and C only
    *e. A, B, and C

4. Which of the following would most likely result in agglutination?
    a. The mixing of similar blood types
    b. The separation of blood cells from plasma
    c. The reaction of antigens to other antigens
    *d. The reaction of antibodies to foreign antigens
    e. The donation of blood to a blood bank

5. A universal receiver can receive a blood transfusion from persons of any blood type. Which blood type(s) would a person need to have to be a universal receiver?
   a. Type A only
   b. Type B only
   *c. Type AB only
   d. Type O only
   e. Either type AB or O

Note: Rh factor compatibility must also be addressed in blood transfusions.

Electronic or Video Portfolios for Students

A teacher may wish to have students document their work and class-related activities in an electronic or video portfolio. The technology exists that can make this effort very student driven. With digital cameras and the ease of use of software, students can create multimedia portfolios that document their work at the same time they expand skills in technology use. Throughout the school year, a visual record of projects, presentations, and group work could be created; this would require access to equipment and software to support the effort. Software to support electronic portfolios is available, and the generation of student Web pages is also a possibility.

The student, teacher, administrator, and parents could view the videos, electronic portfolios, or Web pages periodically during the year to look at student growth. These records could be used to provide feedback and could serve as an excellent record of each student's year. School policies should be followed with regard to confidentiality and anonymity. If a video format were used, a review form could be sent home with the video so when parents and students view the tape together, they can discuss and critique the work represented on the tape. Form 7.1 gives an example.

Structuring Peer Assessment for Cooperative Groups

Cooperative groups are frequently used for work in classes, and peer assessment can help group members assume more personal responsibility for actions within the group. This assessment can include both assessment of each group member by their peers and an assessment of the group as a whole. Forms 7.2 and 7.3 give an example of each.

FORM 7.1
Letter for Video Review

Dear _____ ,
          (parent or guardian)

_____ has created this video of his or her
                 (student's name)
science presentation and would like you to watch the video with him or her. While watching the video with your child, discuss each of the areas included in this letter. Please provide comments for any of the areas. You may use a scale of 1 to 5 (5 being the highest) to rate the science presentation. Focus on the strengths of the performance and also look for areas for improvement.

Preparation: The student was well prepared. She or he knew what to say about the subject and presented the ideas in a meaningful order. The presentation showed evidence of necessary research and planning.

Rating: _____

Comments:

_____

_____

Visuals: The student used pictures or other visuals to clarify the presentation.

Rating: _____

Comments:

_____

_____

Clarity in Speaking: The presentation was clearly delivered and was well paced. The presentation could be heard. The student spoke with expression and showed enthusiasm for the subject.

Rating: _____

Comments:

_____

_____

FORM 7.1 Continued

Eye Contact: During the presentation, the student looked at people and appeared to make eye contact with the audience.

Rating: _____

Comments:

_____

_____

What did you like best about the video and why?

_____

_____

_____

Thank you for your help and interest in your child's education. We are very proud of the work we are doing and wanted to invite you to share some of our experiences.

Sincerely,

_____
(Student's signature)

_____
(Teacher's signature)

FORM 7.2

Peer Assessment for Cooperative Groups
(Integrates Domains)

Group Member Assessed: _____

Assessed by: _____

How well did this group member contribute to the rest of the group?

Action	Excellent—5	4	Acceptable—3	2	Needs to improve—1
Starts task work	Promptly		Needs a little push		Reluctant to start
Stays on task	Almost Always		OK, but could improve		Seldom
Tried to answer own questions	Almost Always		OK but could improve		Seldom
Helps keep noise level down	Almost Always		OK but could improve		Seldom
Shares the work load	Almost Always		OK but could improve		Seldom
Encourages others	Almost Always		OK but could improve		Seldom
Contributes ideas	Almost Always		OK but could improve		Seldom

Score: _____/35 possible

Comments:

FORM 7.3

Group Assessment of Cooperative Roles
(Integrates Domains)

Group Members: _____

How well did your cooperative group do for each of the following?

Action	Excellent—5	4	Acceptable—3	2	Needs to improve—1
Start task work	Promptly		Needs a little push		Reluctant to start
Stay on task	Almost Always		OK but could improve		Seldom
Try to solve or answer your own questions	Almost Always		OK but could improve		Seldom
Keep noise level down	Almost Always		OK but could improve		Seldom
Encourage each other	Almost Always		OK but could improve		Seldom
Divide the work load	Almost Always		OK but could improve		Seldom
Listen to each other's ideas	Almost Always		OK but could improve		Seldom

Score: _____/35 possible

What was the greatest strength of your group in working on this task?

What area in the group work needed most improvement?

Class Assessment of Group Project Presentations

If students are completing presentations based on group work, both the teacher and the members of the class can assess the presentation. The presentation criteria should be shared with students in advance, and details such as the number and kinds of references required to support a project and presentation would need to be communicated to students in the project criteria. The presentation criteria could also be made specific to the topic or concepts being studied. In the sample Forms 7.4 and 7.5, the numbers "4" and "2" are included, in the event that the evaluation of the particular criterion falls in that borderline region.

FORM 7.4

Class Assessment of Group Project Presentation
(Integrates Domains)

Project Topic: _____

Group Names: _____

Presentation Criteria

Criteria	*Excellent—5*	*4*	*Acceptable—3*	*2*	*Needs Improvement—1*
Understanding of the topic	Group knew what they were talking about		For the most part, group knew what they were talking about; a few errors present		Group hesitant at times; numerous errors were present
Questions about the topic	Group could answer all questions		Group could answer most questions		Group unable to answer most questions
Relatedness to unit of study	Project strongly supported unit of study		Project related to unit but lacked some connections		Project seemed vague and unrelated
Presentation skills	High interest; good eye contact; many visual aids; very well organized		Moderate interest; some eye contact; some visual aids; good organization		Low interest; little eye contact; few visual aids; poorly organized
Research references	5 or more references; both primary & secondary sources used		3-4 references; limited variety in types of sources		Fewer than 3 references

Score: _____/30 possible

What was one strength of the group presentation?

What is one suggestion you would make to improve the presentation?

Enger, S., & Yager, R. *Assessing Student Understanding in Science.* © 2001 by Corwin Press, Inc.

FORM 7.5

Student Self-Assessment of Role in Group Project
(Integrates Domains)

Project Topic: _____

Name: _____

Criteria for Self-Assessment

Criteria	Excellent—5	4	Acceptable—3	2	Needs Improvement—1
What I learned	I learned many things about the topic.		I learned some things about the topic.		I didn't learn much about the topic.
My understanding of the concepts	I understand the concepts well enough to explain them to others.		I understand the concepts fairly well but still have some questions.		I have quite a lot of difficulty understanding the concepts.
Finding references & resources	I was able to find the number requested with little or no difficulty.		I had some difficulty finding these.		I was not able to find these.
My contributions	I often helped others with their tasks.		I occasionally helped others with their tasks.		I only had time to work on my own selected tasks.
Social skills	My group worked together very well.		My group had a few problems, but usually we worked together quite well.		My group did not work together very well.
Doing my tasks	I completed all of my tasks on time.		I completed most of my tasks on time.		Some of my task work remained uncompleted.

Score: _____/30 possible

My greatest contribution to the group was _____.

The next time I work with a group, I will try to improve my _____.

8

Assessment Examples for Grades 9 Through 12

In this chapter, you can find suggestions and examples of assessments that address

- Science processes

- Understanding selected electricity concepts

- Understanding temperature and heat

- Understanding environmental sciences

- Food chains and food webs

- The student's view of the way science happens

- Understanding the nature of science

Using the Student Laboratory Environment Inventory

Laboratory teaching is an integral part of the science classroom. Although much is known about hands-on learning and many studies have corroborated that learning occurs best when one is doing, how can the effects of laboratory instruction on student learning and attitudes be measured? As mentioned in Chapter 5, Fraser, Giddings, and McRobbie (1992) designed the SLEI for situations in which a separate laboratory class exists. As noted earlier, the SLEI integrates domains, consists of

35 items that are designed to measure five different dimensions, and uses the response alternatives of *almost never, seldom, sometimes, often,* and *very often.* Two SLEI forms can be used: An Actual Form asks students to note what actually takes place in the laboratory classroom, whereas the Preferred Form asks students to respond to what they would prefer to have take place in the laboratory class.

Looking at Science Processes

Students may be using one or several process skills during an instructional period, and the teacher could use a predetermined observation checklist to document the use of process skills addressed in a particular class activity. Examples of items that could be used to assess some of the processes of science are provided in Form 8.1. Students should be informed about the different performance levels and what would be considered an indicator of being able to use this process skill.

Assessing Specific Areas of Understanding

Figures 8.1 and 8.2 and Form 8.2 are examples of assessment instruments pertaining to students' understanding of electricity, temperature and heat, and environmental science, respectively.

📖 *Applying Concepts of Heat:*
Preassessment for Concepts of Heat Transfer
(Application Domain)

Ask students to each draw a picture of an ice cube in a glass of water. Students should then draw arrows to show the directions of heat flow. Students should also provide explanations to support their drawings.

Next, ask students to each draw a picture of a lighted candle. Students should then draw arrows to show the directions of heat flow. Students should also provide explanations to support their drawings.

The instructor could set up a glass of water and also a lighted candle as models. After students have completed their drawing and writing, they could compare and contrast their answers, and the instructor could note misunderstandings on which to focus instruction. A postassessment could ask students to respond to these same situations. If students' drawings are saved, these could be returned to the students, and the students could be asked to revise and add to the drawings and descriptions as necessary.

(continued on p. 155)

FORM 8.1

Observation Checklist

Activity Description:

Student: _____

Rating Scale

PP	LP	NMI	NA, NO
Proficient performance	Limited proficiency: Can use but needs more practice	Cannot do or use: Needs more instruction and practice	Not applicable, not observed

Circle the processes that apply to the activity:

Observing   Classifying   Communicating   Measuring

Predicting   Inferring   Identifying and controlling variables

Formulating and testing hypotheses   Interpreting data

Defining operationally   Experimenting   Constructing models

1. Appropriate process skills were used while conducting the activity. Notes:	PP	LP	NMI	NA, NO
2. Appropriate process skills were used to determine the outcome of the activity. Notes:	PP	LP	NMI	NA, NO
3. Work was recorded in the learning or laboratory log. Notes:	PP	LP	NMI	NA, NO
4. Additional experiences are needed to practice these skills. Notes:				

Recommendations:

FIGURE 8.1.

Understanding Electricity (Concept Domain)

For each question, an asterisk (*) indicates the best answer.

1. An electrical current is caused by moving electrons. Which of the following is true about the charge on an electron?
   a. Electrons have a positive charge.
   *b. Electrons have a negative charge.
   c. Electrons are neutral.
   d. Electrons can have either a positive or negative charge.

2. The quantity of charge that passes through a circuit is measured in
   a. Amps
   b. Volts
   *c. Coulombs
   d. Watts

3. The amount of energy dissipated each second in a circuit depends on which of the following?
   a. The voltage only
   b. The current only
   c. The resistance only
   *d. The current and the voltage

Use the following information to answer questions 4 through 7.

Four different kinds of cells or batteries could be created with the following materials:
   a. Zinc and carbon
   b. Lead and acid
   c. Nickel and cadmium
   d. Mercury

For each of the following devices, select the most suitable cell or battery to use
and write the letter in the blank.

_____ 4. A hearing aid (d)
_____ 5. A model car (a)
_____ 6. A real car (b)
_____ 7. A rechargeable electric shaver (c)

8. Draw circuit diagrams to illustrate each of the following situations:
   a. A battery made of three cells is used to supply energy for two bulbs connected in parallel.
   b. A single cell is used to supply energy to two bulbs in series. A voltmeter is connected across one of the bulbs to measure the potential drop across the bulb.

FORM 8.2

Preassessment of Understanding of Temperature and Heat
(Concept Domain)

Check whether each statement is true or false. If you are not certain if the statement is true or false, check "I'm not sure."

1. Temperature is the quantity of heat that is absorbed by an object.

   True _____   False _____   I'm not sure _____

2. When heat is absorbed by an object, its temperature always increases.

   True _____   False _____   I'm not sure _____

3. When heat is given off by an object, its temperature always decreases.

   True _____   False _____   I'm not sure _____

4. Temperature is a physical property of a substance. Some substances are cold, and some are warmer.

   True _____   False _____   I'm not sure _____

5. Objects left inside a container for a long time will all reach the same temperature.

   True _____   False _____   I'm not sure _____

6. If the temperature of an object that is heated remains the same, then there is probably a change in the state of the object.

   True _____   False _____   I'm not sure _____

7. A larger piece of ice has a lower temperature than a smaller one; therefore, it melts more slowly.

   True _____   False _____   I'm not sure _____

8. Temperature is a kind of energy. Objects with higher temperatures contain more heat, and objects with lower temperatures contain less heat.

   True _____   False _____   I'm not sure _____

Please write down other things that you know or do not know about temperature or heat.

SOURCE: Adapted from Guo (1993).

Enger, S., & Yager, R. *Assessing Student Understanding in Science.* © 2001 by Corwin Press, Inc.

FIGURE 8.2.

___

Environmental Science Concepts (Concept Domain)

Select the best answer.

1. A community is made up of clover, rabbits, and foxes. If an extended drought occurred in the area, the most immediate effect would likely be

   a. A decrease in the number of foxes
   b. A decrease in the number of rabbits
   *c. A decrease in the amount of clover
   d. An increase in the number of rabbits

2. The number of different species of birds, insects, and mammals occupying a given ecosystem is primarily dependent on which of the following?

   a. The amount of competition between species
   *b. The diversity of vegetation in the system
   c. The absence of large predators
   d. The time when vegetation is the thickest

3. Identify the type of area that would have these plants and animals: caribou, lemmings, lichens, moss, short grass, wolves.

   *a. The North American tundra
   b. The eastern deciduous forest
   c. The southwestern desert
   d. A tropical rain forest

4. If the producers in an ecosystem were suddenly unable to use radiant energy from the sun, what of the following would immediately happen?

   a. Respiration in producers would suddenly cease.
   *b. Photosynthetic activity would stop.
   c. The system would increase its biomass.
   d. The decomposers in the system would cease to function.

5. What statement best describes the relationship between the number of producers in an ecosystem and the number of primary consumers in the same system?

   *a. An increase in the number of producers is usually accompanied by an increase in primary consumers.
   b. An increase in primary consumers is usually accompanied by an increase in producers.
   c. The number of producers in an ecosystem and the number of primary consumers in the same system are not related.
   d. The number of producers depends on the number of primary consumers.

SOURCE: Adapted from Fleetwood (1972).

FIGURE 8.3.

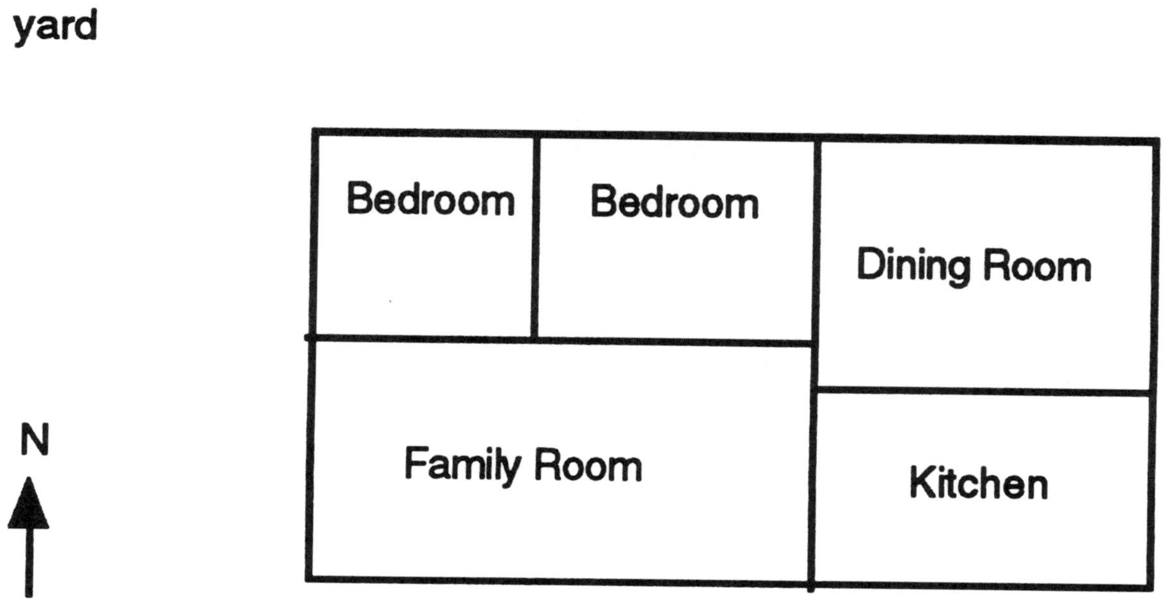

yard

Energy Efficiency Assessment 📖
(Application Domain)

A basic floor plan for a house is set out (see Figure 8.3), and although you plan to have air conditioning, you want to keep your utility bills as low as possible. Show where you would locate trees around the yard to help conserve energy. Assume that the house is in an area where you can have snow in the winter and temperatures as high as in the 90s Fahrenheit in summer. Explain why you would place the trees where you did.

Food Chains and Food Webs 📖
(Application Domain)

When one organism becomes food for another organism and energy is transferred via eating and being eaten, this flow of energy is called a *food chain*. This chain began when the producers absorbed the sun's energy to produce their own food. Then consumers ate the producers to obtain their energy, and consumers also may have been eaten by other consumers.

FIGURE 8.4. A Food Web

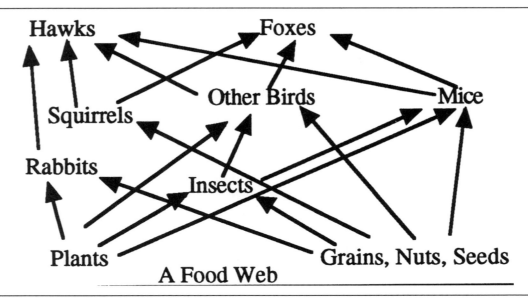

The following diagram is an example of a food chain:

Plants ⟶ Insects ⟶ Mouse ⟶ Hawk

Eating and being eaten becomes a much more complex process in the ecosystem, and food chains begin to interconnect to form *food webs*, as in the example shown in Figure 8.4.

Using the information presented about food chains and food webs, answer the questions that follow.

1. When a large group of cats was introduced to the food web shown, almost all of the mice were eaten in just a few weeks. Which organisms would be most likely to have the largest increase in population as a result?
   *a. The insects
   b. The foxes
   c. The rabbits
   d. The hawks
   e. The squirrels

2. The decomposers, organisms that obtain their food from both wastes and dead organisms, are not shown in the food web. Decomposers are important in the food web because they do which of the following?
   *a. Recycle nutrients
   b. Inhibit diseases
   c. Control population growth

      d. Compete with producers
      e. Compete with consumers

3. Actually, the food web shown is a hypothetical example. In reality, it could be expected that food webs would be
      a. Very similar to the illustration.
    *b. More complex than the illustration.
      c. Simpler than the illustration.
      d. Uncommon in nature.
      e. Nonexistent in nature.

4. Based on the information provided, which of the following statements is *not* correct?
      a. Each level of the food chain depends on the level below.
    *b. Energy travels in both directions in a food chain.
      c. The food web shows many interrelated food chains.
      d. Consumers often compete for the same food source.
      e. Most consumers eat a variety of foods.

5. Which of the following would be a producer in the ocean?
      a. Shrimp
      b. Mussels
    *c. Algae
      d. Salt
      e. Whales

6. In the food chain, consumers are ranked according to
      a. Their size.
      b. Their weight.
      c. Where they live.
    *d. What they eat.
      e. How much they eat.

Assessing Students' Views of the Scientific Process

Figure 8.5 and Form 8.3 are examples of student self-assessment instruments regarding the Nature of Science domain.

FIGURE 8.5. Your View of the Way Science Happens (Nature of Science Domain)

Student Name: _____

For each question, mark the letter that indicates your view of the best response.

1. The main reason why scientists study plants is
   A. To teach farmers how to grow more food
   B. To learn how to make better medicine from plants
   *C. To be able to explain how plants grow
   D. To be able to tell what is the best soil for plants

2. Over 200 years ago, a scientist, Sir Isaac Newton explained the motion of planets with what is called gravity. Scientists decided that Newton's explanation was right. Not long ago, a scientist, Albert Einstein, said that his idea (relativity) could explain everything that Newton's idea (gravity) could explain, plus more. Today, scientists accept Einstein's idea (relativity). What do they say about Newton's idea?
   A. Newton's idea was wrong, but Newton didn't study as many things as Einstein did
   *B. Newton's idea was right, but Newton's idea can't explain as much as Einstein's idea can
   C. Newton's idea will only work if you are on a planet somewhere else in space
   D. Newton's idea is better than Einstein's because it has been used longer

3. A scientist in Australia says that she has seen signs of plant life on the planet Venus. American scientists will believe this scientist
   *A. If other scientists also see these signs of plant life
   B. If she can tell them what type of plants these probably are
   C. If the Australian government will stand behind the scientist's work
   D. If other scientists agree that there is oxygen on Venus

4. If we ask an astronomer to explain why some stars look brighter than others, her/his explanation would
   A. Tell us why it is necessary for stars to differ in brightness
   *B. Use scientific laws
   C. Be made with mathematical equations and formulas
   D. Use observations and data from astronomy

5. Science is an attempt to
   A. Make sure that what has been discovered about the world is really true
   *B. Give us laws and theories that help explain nature
   C. Discover, collect, and group facts about nature
   D. Find ways to make people's lives better

6. Which of these statements best describes scientific knowledge?
   A. Scientific knowledge is a well-organized collection of facts.
   *B. Today's scientific knowledge builds on data and ideas from the past.
   C. Today's scientific knowledge has been produced by today's scientists.
   D. Scientific knowledge contains only statements that are 100% true.

7. Are biology, chemistry, and physics related to each other?
   A. They are not related, because they are built on different basic ideas
   *B. They are related because some of the ideas in each relates to the ideas in the others
   C. They are related because mathematics ties them together
   D. They are not related because each one studies something very different from what is studied by the others

FIGURE 8.5. Continued

8. Today, scientists are doing experiments to see if one of Einstein's theories is right when it predicts that light rays will bend as they pass near a very large object, such as a star. This is an example of
   *A. How theories suggest experiments to try
   B. How it is important to have an exact measurement of how fast light travels
   C. How experiments must be done to prove that a theory is really true
   D. How theories are still doubted long after they are proven true

9. Designing a television set is a problem for
   A. Science, because to design a set, you need brain power
   B. Science, because to design a set, you need to do experiments
   *C. Technology, because it leads to a useful device
   D. Technology, because it involves working with electricity

10. Betty is planning an experiment to find out whether potassium is important in the growth of a certain plant. Her teacher suggests growing one group of these plants in soil with nitrogen and phosphorus but no potassium. The teacher will probably also suggest that Betty grow another group in soil with
    A. Potassium only
    *B. Nitrogen, phosphorous, and potassium
    C. Nitrogen and potassium but with no phosphorus
    D. Nitrogen and phosphorus but with no potassium

11. All of the following statements deal with how science and technology are connected. Which statement best describes the connection?
    *A. Technology uses scientific knowledge to help solve practical problems
    B. Science depends on technology for ideas and for help in planning experiments
    C. The laws used in science come from technology
    D. Technology is the part of science that solves mechanical problems

12. John was asked to describe a scientific law. Which of the following descriptions should he give?
    A. A scientific law is an exact report of what scientists observe
    *B. A scientific law says how one type of event in nature is related to another event
    C. A scientific law is an explanation of an event in nature, and it uses things that can't be seen
    D. A scientific law is a rule that nature makes, and it cannot be broken

13. A certain law cannot explain some of the facts that it should explain. What may scientists do with this law?
    A. Scientists may change the unexplained facts so that the law can explain them
    *B. Scientists may change the law so that more of the facts can be explained
    C. Scientists should throw out the law and make a new one immediately
    D. Scientists should show that the law is wrong in the case of all facts

14. Robert Hooke did many experiments with springs and found that most springs stretch by twice as much when the amount of weight hung on the spring is made twice as great. Springs stretch by three times when the amount of weight is made three times as great, and stretch four times as much when four times as much weight, and so on. This led Hooke to make a general rule for how much springs stretch when different weights are hung from them. This story is an example of how some scientists
    A. Have come up with a scientific theory
    B. Have tested a scientific hypothesis
    *C. Have come up with a scientific law
    D. Have come up with ideas about real things from ideas about imaginary things

(continued)

FIGURE 8.5. Continued

15. When a new theory is suggested, scientists will probably decide that it is a good theory
    A. If they think the theory is true
    B. If the theory can be put into a mathematical equation
    C. If the theory fits with what they observe and agrees with their other ideas
    *D. If the theory fits with observations

16. When scientists decide that a new theory is an acceptable theory, we can say that
    *A. Science's ideas and explanations of nature have increased.
    B. One more of the laws of nature is now known.
    C. We are closer to the end of the search for scientific knowledge.
    D. Science has discovered new experimental evidence.

17. Many people say that scientists behave in special ways. For example, they observe carefully, they do not jump to conclusions, and they are very exact. If we wanted to see scientists behave this way, it would be best to watch them when
    *A. They are doing experiments
    B. They are doing work outside of science
    C. They are doing almost anything
    D. They are with their family and friends

18. In discussing the problem of nuclear weapons, a famous scientist said that we must keep on doing experiments with nuclear weapons. The scientist is probably
    A. Right, because his or her scientific attitude makes the answers to problems more correct
    B. Wrong, because scientists seem to be trying to destroy the world
    C. Right, because scientific results are right more often than other kinds of results
    *D. No more right or wrong than any other intelligent person, unless he or she studied the problem

19. A book about science says, "Scientists do experiments to ask nature questions." This means that experiments are used in science
    A. To prove that nature follows rules
    B. To learn by trying different solutions to a problem until one works
    *C. To see if predictions made from scientists' ideas are right
    D. To try to find out about the origin of humans

20. Lisa has a good imagination, but she may never become a scientist because
    A. She would not want to give up being able to think anything that she wants to think
    B. People with a good imagination usually become artists and writers
    *C. She might like some other field better than science
    D. Science has too many facts

21. When scientists carefully measure any quantity many times, they expect that
    A. All of the measurements will be exactly the same
    B. Only two of the measurements will be exactly the same
    C. All but one of the measurements will be exactly the same
    *D. Most of the measurements will be close but not exactly the same

22. Which of the following is the description of a scientific theory?
    A. A scientific theory uses arithmetic.
    B. A scientific theory describes a scientist.
    C. A scientific theory describes an experiment.
    *D. A scientific theory helps explain why some things act the way they do.

SOURCE: Adapted from Cooley and Klopfer (1961).

FORM 8.3

What You Think About the Nature of Science
(Nature of Science Domain)

Student Name: _____

For each question, circle the letter that best indicates your viewpoint.

	Strongly Agree	Agree	Disagree	Strongly Disagree
1. Science means questioning, explaining, and testing.	a	b	c	d
2. Science means studying the concepts developed and known by scientists.	a	b	c	d
3. Science means working with various objects and materials in classrooms and laboratories.	a	b	c	d
4. Science deals with activities that affect living: in home, schools, communities, and nations.	a	b	c	d
5. Science is a human activity that involves acting on questions about the universe.	a	b	c	d
6. Science is a body of knowledge, developed over years, about the universe.	a	b	c	d
7. Science can be defined by what scientists know.	a	b	c	d
8. Science is an attempt to know more about the world around us.	a	b	c	d
9. Science is a way of viewing the universe and how it works.	a	b	c	d
10. Science is based on curiosity about objects and events in the universe.	a	b	c	d
11. Science is based on attempts to answer the questions about the objects and events in the universe.	a	b	c	d
12. Science must include tests in nature that illustrate the validity of personal explanations offered.	a	b	c	d
13. Science is a self-correcting human endeavor.	a	b	c	d
14. Any theory or concept of science can be challenged.	a	b	c	d

Enger, S., & Yager, R. *Assessing Student Understanding in Science.* © 2001 by Corwin Press, Inc.

Glossary of Assessment-Related Terminology

Assessment

The key questions are, What do students know? How well do teachers teach? Do our schools work? (Kulm & Malcom, 1991).

Alternative assessment applies to assessments that differ from the multiple-choice, timed, one-shot approaches that characterize most standardized and many classroom assessments (Marzano et al., 1993). The reason to promote using alternative assessment is to avoid the drawback of ignoring other performances in addition to outcomes coming from standardized tests.

Authentic assessment conveys the idea that assessment should engage students in applying knowledge and skills in the way they are used in the real world. The key point is, "test students in context" rather than by using the standardized tests. It also reflects good instructional practice, so that teaching to the test is desirable (Wiggins, 1989a, 1989b).

Holistic assessment encompasses the totality of student performance in the classroom, and like performance assessment and beyond test outcomes, many components of student learning are of importance. An excellent way for a teacher to evaluate student work in a classroom or laboratory is to use an observational rubric to record observations and comments about student performance. Observations help the teacher note not only when students are successful, but observations can help identify students who may have difficulty with equipment usage, those who contribute ideas, and those who hesitate to participate (Raizen & Kaser, 1989).

Performance assessment refers to the variety of tasks and situations in which students are given opportunities to demonstrate their understanding and to thoughtfully apply knowledge, skills, and habits of mind in a variety of contexts (Marzano et al., 1993).

Application Domain

The key question is, To what extent students can transfer and use effectively what they have learned in a new situation, especially in daily life (Gronlund, 1988)? STS stresses the association of scientific knowledge with students' social and living experiences by using current social issues to help students see their connections. Starting from the concerns in the real world may be a way to diminish the learning gap between the two worlds: the school-science experiences and the personal technology experiences (Yager & McCormack, 1989).

Attitude Domain

This is a state of mind or a feeling. The term is very broadly used in discussions about science education but often with various meanings. It is possible to distinguished two broad categories: attitude toward science (e.g., interest in science, attitude toward scientists, or attitudes toward social responsibility in science) and scientific attitude, (i.e., open-minded, honesty, or skepticism; Gardner, 1975). Whereas *Science for All Americans* (AAAS, 1990) postulates scientific literacy for all Americans, STS curriculum tries to promote students' positive attitude toward science, which includes an I-can-do-it attitude and decision making about social and environmental issues (Yager & McCormack, 1989).

Behaviorist Teaching Approach

In the behaviorist approach to teaching, Mayer (1987) noted that what is of interest to the teacher is the relationship between instructional manipulations and the outcome performance. Behaviorist approaches tend to focus more on overt behaviors, where cognitive approaches focus on both overt and covert behaviors.

Cognitive Teaching Approach

The cognitive approach attempts to understand (a) how instructional manipulations affect internal cognitive processes, such as paying attention; (b) how these processes result in the acquisition of new knowledge; and (c) how new knowledge influences performance, such as on tests. The goal is to explain the relation between stimulus and response by describing the intervening cognitive processes and structures (Mayer, 1987).

Concept Domain

Conceptual systems are primarily structured via *kind* or *is-a* hierarchies (i.e., Tweety is a canary, which is a kind of bird, which is a kind of animal, which is a kind of thing) and *part-whole* hierarchies (i.e., a toe is part of a foot, which is part of a leg, which is part of a body; Thagard, 1992). Helping students to construct the natural world into their own categorization in terms of scientific knowledge is one of the basic functions of science education. This domain includes facts, laws (principles), theories, and internalized knowledge of students (Yager & McCormack, 1989).

Concept Mapping

Concept mapping is a technique for externalizing concepts and propositions. *Propositions* are two or more concept labels linked by words in a semantic unit. Concept maps are an explicit, overt representation of the concepts and propositions a person holds. They allow teachers and learners to exchange views on why a particular prepositional linkage is good or valid or to recognize missing linkages between concepts (Novak & Gowin, 1984). It is always used with interviews to sort out the underlying meaning of each proposition.

Constructivism

The learner constructs and frames meaning in terms of his or her existent experiences in a social context. Knowledge is not passively received but is actively built on past experiences. Cognition is adaptive and helps the learner organize the experiential world, not the discovery of ontological reality. The learner does not find truth but instead constructs viable explanations based on personal experiences (Wheatley, 1991).

Cooperative Learning

It could be argued that students should learn through discussion in social contexts and that personal learning cannot be independent from external social activities. Working in small groups can provide students opportunities to challenge one another's ideas, and this can lead them to realize the need to reorganize and reconceptualize their thinking. Johnson & Johnson (1990) provided evidence that when students learn more social skills, their performance on standardized math computation scales became better.

Creativity Domain

Torrance (as cited in Penick, 1996) described creativity as a process of becoming sensitive to problems, deficiencies, gaps in knowledge, missing elements, and disharmonies. Creativity involves the identification of the problematic; the search for solutions, the generation of guesses, or the formulation of hypotheses about the deficiencies; testing and retesting the hypotheses and the potential modification and retesting; and last, the communication of the results. STS teaching strategies stress the role of guessing as a basic element of creativity. When students are viewing natural phenomena, to stimulate their creativity and interpretation of results, they should be encouraged to present wild guesses to generate diverse perspectives (Yager & McCormack, 1989).

STS teaching strategies stress the role of *guessing* as the basic element of creativity and suggests that students give wild guesses to serve as a diversity of perspectives in seeing the same natural phenomena to stimulate students' creativity and interpretation of results (Yager & McCormack, 1989).

Evaluation

Evaluation is a systematic process of collecting, analyzing, and interpreting information for judging decision alternatives. It not only uses quantitative measurements and, in some cases, qualitative information but also involves value judgments. In the classroom, evaluation is the process of interpreting students' performances for both formative and summative purposes.

Informal Test

An informal test is one constructed by a teacher for use in one or a limited number of occasions where comparability or results across groups is not essential.

Law

In science, laws are often expressed as equations relating measurable parameters. Laws indicate the mathematical relations that hold true among quantities of certain empirical parameters. Laws provide us with knowledge of facts but not with an explanation as to why the facts are as they are (Dilworth, 1981).

Measurement

Measurement is the process of assigning numbers to attributes of individuals according to specified rules (it always refers to quantification).

Nature of Science Domain

STS stresses the nature of the development of scientific knowledge and the influence coming from the external social factors, which in turn discount the absolute objectivity of science. Abimbola (1983) identified these basic themes about the nature of science: Observations are theory laden, and the existing paradigms determine problem selection, instrumentation, and the inferential techniques and models used. The scientific community ultimately drives the choice of the scientific question, and formal logic is replaced by a reliance on the detailed study of the history of science. Continued research and the accompanying critique by the scientific community are at the core of science. Science progresses in two phases: normal science and revolutionary science. Normal science operates in the context of a shared paradigm and is responsible for generating scientific revolutions. The more important events in the history of science are those revolutions that change paradigms. Progress in science is, therefore, noncumulative because of paradigm shifts. Observational data do not remain the same from one scientific revolution to another because scientific paradigms are incommensurable, and scientists view events through the lens of a new paradigm.

As Lederman (1992) suggested, students' conception of the nature of science was inadequate, and Carey, Evans, Honda, Jay, and Unger (1989) suggested that the teaching of the nature of science can improve learning. Other researchers (e.g., Matthews, 1991) have suggested that the use of historical case studies may be one way to build greater student understanding about the nature of science.

Process Domain

Practical work, including hands-on activities, scientific inquiries, or experiments, is always cited as the most powerful approach to helping students understand scientific knowledge. At the same time, Hodson (1992) argued that the skills-based approach of practical work was philosophically unsound, educationally worthless, and pedagogically dangerous. STS suggests a holistic view of assessment to promote valid process learning, especially emphasizing the role of creativity in the data analysis process (Yager & McCormack, 1989).

Reliability

Reliability is the consistency or the degree of consistency between two or more measures of the same thing. The key questions are those of (a) Are the performances of students consistent? and (b) To what degree?" High reliability of a test does not necessarily indicate a good measurement unless there is also high validity.

Rubrics, Scoring Guides

Rubrics, or scoring guides, are used to score nontraditional assessments. A typical rubric (a) contains anywhere from 3 to 10 levels of performance, (b) states all the major dimensions to be assessed, and (c) provides necessary information about idiosyncrasies in the question or equipment. Holistic, analytic, or component rubrics can be used.

Scientific Inquiry

Scientific inquiry is the process in which phenomena are investigated and the results of the observations are interpreted. In a positivist viewpoint, students can reach scientific knowledge without help from external guidance. Yet more and more research confirms that students' prior knowledge may bias what they observe and create misconceptions (Matthews, 1988). According to Ausubel (1968), the most important single factor influencing learning is what the learner already knows. If this can be ascertained, then the instruction should reflect this.

Standardized Test

A standardized test is typically constructed for use in more than one setting. It is standardized in the sense that the administrative procedures, directions, apparatus, and scoring are fixed by the constructors so that the test may be administered and scored identically by different examiners in different settings to achieve comparable results across all examined.

Test

A *test* is a systematic procedure for observing behavior and describing it with the aid of numerical scales or developed categories.

Theory

The primary function of a theory is to provide the conception of a mechanism that can explain the empirical regularities behind appearances, as do laws. Yet theories simply offer potentially true descriptions of a reality. Thus, a theory can one day come to be abandoned as a result of measurement made by ever more sensitive instruments or the arrival of a superior alternative (Dilworth, 1981).

Validation

Validation is "the process by which a test user collects evidence to support the types of inferences that are to be drawn from test scores (Crocker & Algina, 1986, p. 217).

Validity

"The degree to which the test actually measures what it purports to measure" (Anastasi, 1982, p. 28) is called *validity*. The key questions are (a) Did we test what we wanted to test? and (b) To what degree? High validity usually indicates high reliability.

References

Abimbola, I. O. (1983). The relevance of the "new" philosophy of science for the science curriculum. *School Science and Mathematics, 83*(3), 182-90.

Aikenhead, G. S. (1973). The measurement of high school students' knowledge about science and scientists. *Science Education, 57*(4), 539-49.

Aikenhead, G. S. (1979). Science: A way of knowing. *The Science Teacher, 46*(6), 23-25.

Akindehein, F. (1988). Effect of an instructional package on pre-service science teachers' understanding of the nature of science and acquisition of science-related attitudes. *Science Education, 72*(1), 73-82.

American Association for the Advancement of Science. (1968). *Science: A process approach.* Washington, DC: Author.

American Association for the Advancement of Science. (1990). *Science for all Americans.* Washington, DC: Author.

American Association for the Advancement of Science. (1993). *Benchmarks for science literacy.* Washington, DC: Author.

Anastasi, A. (1982). *Psychological testing.* New York: Macmillan.

Angelo, T. A., & Cross, K. P. (1993). *Classroom assessment techniques: A handbook for college teachers.* San Francisco: Jossey-Bass.

Association for Science Education. (1986). *STS examination items for science in a social context.* Oxford, England: Basil Blackwell.

Ausubel, D. P. (1968). *Educational psychology: A cognitive view.* New York: Holt, Rinehart & Winston.

Barenholz, H., & Tamir, P. (1992). A comprehensive use of concept mapping in design instruction and assessment. *Research in Science and Technological Education, 10*(1), 37-52.

Barnum, C. R. (1996). How do students really view science and scientists. *Science and Children, 34*(1), 30-33.

Barnum, C. R. (1997). Students' view of scientists and science: Results from a national study. *Science and Children, 354*(1), 18-24.

Barufaldi, J. P., Bethel, L. J., & Lamb, W. G. (1977). The effect of a science methods course on the philosophical view of science among elementary education majors. *Journal of Research in Science Teaching, 14*(4), 289-294.

Brooks, J. G., & Brooks, M. G. (1993). *In search of understanding: The case for constructivist classrooms.* Alexandria, VA: Association for Supervision and Curriculum Development.

Burry-Stock, J. A. (1993). *Expert science teaching evaluation model (ESTEEM) training manual.* Kalamazoo, MI: Western Michigan University, The Evaluation Center.

Carey, S., Evans, R., Honda, M., Jay, E., & Unger, C. (1989). An experiment is when you try it and see if it works: A study of grade 7 students' understanding of the construction of scientific knowledge. *International Journal of Science Education, 11,* 514-529.

Champagne, A. B., & Newell, S. T. (1992). Directions for research and development: Alternative methods of assessing scientific literacy. *Journal of Research in Science Teaching, 29*(8), 841-860.

Cooley, W. W., & Klopfer, L. E. (1961). *Test on understanding science.* Princeton, NJ: Educational Testing Service.

Crocker, L., & Algina, J. (1986). *Introduction to classical and modern test theory.* New York: CBS College Publishing.

Csikszentmihalyi, M. (1990). *Flow: The psychology of optimal life experience.* New York: HarperCollins.

Csikszentmihalyi, M. (1996). *Creativity: Flow and the psychology of discovery and invention.* New York: HarperCollins.

Dilworth, C. (1981). *Scientific progress.* Dordrecht, Holland: Dordrecht Publishers.

Enger, S. K. (1997). *The relationship between science learning opportunities and ninth grade students' performance on a set of open-ended science questions.* Unpublished doctoral dissertation, The University of Iowa, Iowa City.

Felker, E. (1974). *Building positive self-concepts.* Minneapolis, MN: Burgess.

Fleetwood, G. R. (1972). *Environmental science test.* Raleigh, NC: North Carolina Department of Public Instruction.

Fort, D. C., & Varney, H. L. (1989). How do students see scientists: Mostly male, mostly white, and mostly benevolent. *Science and Children, 26*(8), 8-13.

Fraser, B. J., Giddings, G. J., & McRobbie, C. J. (1992). Assessing the climate of science laboratory classes. *What Research Says to the Science and Mathematics Teacher, 8,* 1-8.

Gardner, P. L. (1975). Attitudes to science: A review. *Studies in Science Education, 2,* 1-41.

Gay, L. R., & Airaisian, P. (2000). *Educational research: Competencies for analysis and application.* Upper Saddle River, NJ: Prentice Hall.

General Educational Development. (1990). *The 1988 tests of general educational development: A preview.* Washington, DC: American Council on Education.

Giddings, G. (1993). *Student instruction and motivation survey.* Perth, West Australia: Curtin University.

Gronlund, N. E. (1988). *How to construct achievement tests.* Englewood Cliffs, NJ: Prentice Hall.

Gronlund, N. E., & Linn, R. L. (1990). *Measurement and evaluation in teaching.* New York: Macmillan.

Guo, C. J. (1993). *Alternative frameworks of motion, force, heat, and temperature: A summary of studies for students in Taiwan.* Paper presented at the annual meeting of the National Association for Research in Science Teaching, Atlanta, GA.

Hodson, D. (1992). Assessment of practical work: Some considerations in philosophy of science. *Science & Education, 1*(2), 115-144.

Hodson, D., & Reid, D. J. (1988). Science for all: Motives, meanings and implications. *The School Science Review, 69*(249), 653-661.

Hopkins, D. (1993). *A teacher's guide to classroom research.* Buckingham, UK: Open University Press.

Jellen, H. G., & Urban, K. K. (1986). The TCT-DP (test for creative thinking-drawing production): An instrument that can be applied to most age and ability groups. *The Creative Child and Adult Quarterly, 11*(3), 138-144.

Johnson, D. W., & Johnson, R. T. (1983). Interdependence and interpersonal attraction among heterogeneous and homogeneous individuals: A theoretical formulation and a meta-analysis of the research. *Review of Educational Research, 53*(1), 5-54.

Johnson, D. W., & Johnson, R. T. (1990). Social skills for successful group work. *Educational Leadership, 47*(4), 29-33.

Knorr-Cetina, K. D. (1981). *The manufacture of knowledge: An essay on the constructivist and contextual nature of science.* New York: Pergamon.

Kulm, G., & Malcom, S. M. (1991). *Science assessment in the service of reform.* Washington, DC: American Association for the Advancement of Science.

Kuhn, T. S. (1962). *The structure of scientific revolutions.* Chicago: University of Chicago Press.

Lederman, N. G. (1992). Students' and teachers' conceptions of the nature of science: A review of the research. *Journal of Research in Science Teaching, 29,* 331-359.

Marzano, R. J., Pickering, D., & McTighe, J. (1993). *Assessing student outcomes: Performance assessment using the dimensions of learning model.* Alexandria, VA: Association for Supervision and Curriculum Development.

Matthews, M. R. (1988). A role for history and philosophy in science teaching. *Educational Philosophy and Theory, 20*(2), 67-75.

Matthews, M. R. (1991). *History, philosophy and science teaching.* New York: Teachers College Press.

Matthews, M. R. (1994). *Science teaching: The role of history and philosophy of science.* New York: Routledge.

Mayer, R. E. (1987). *Educational psychology.* Dubuque, IA: Little, Brown.

Millar, R. (1989). Constructive criticisms. *International Journal of Science Education, 1,* 587-596.

National Assessment of Educational Progress. (1978). *The third assessment of science (1976-1977).* Denver, CO: Author.

National Council on Measurement in Education. (1995). *Code for professional responsibilities in educational measurement (CPR).* Washington, DC: Author.

National Research Council. (1996). *National science education standards.* Washington, DC: National Academy Press.

National Science Teachers Association. (1982), *An NSTA position statement.* Washington, DC: Author.

Nersessian, N. (1989). Conceptual changes in science and in science education. *Synthese, 80*(2), 163-183.

Nitko, A. J. (1996). *Educational assessment of students.* Englewood Cliffs, NJ: Prentice Hall.

Novak, J. D. (1981). Applying learning psychology and history of science to biology teaching. *The American Biology Teacher, 43*(1), 12-20, 42.

Novak, J. D., & Gowin, D. B. (1984). *Learning how to learn.* New York: Cambridge University Press.

Page, E. (1958). *Teacher comments and student performance.* Englewood Cliffs, NJ: Prentice Hall.

Penick, J. E. (1996). Creativity and the value of questions in STS. In R. E. Yager (Ed.), *Science/technology/society as reform in science education* (pp. 84-94). Albany: State University of New York Press.

Pierce, L. V., & O'Malley, J. M. (1992). *Performance and portfolio assessment for language minority students.* Washington, DC: National Clearinghouse for Bilingual Education.

Raizen, S. A., & Kaser, J. S. (1989). Assessing science learning in the elementary school: Why, what, and how? *Phi Delta Kappan, 70,* 718-722.

Shavelson, R., Baxter, G., & Pine, J. (1992). Performance assessment: Political rhetoric and measurement reality. *Educational Researcher, 21*(4), 22-27.

Stiggins, R. J. (1994). *Student-centered classroom assessment.* New York: Macmillan.

Strauss, S., & Stavey, R. (1983). *Educational developmental psychology and curriculum development: The case of heat and temperature.* Ithaca, NY: Cornell University, International Seminar on Misconceptions in Science and Mathematics.

Swindoll, C. R. (1994). *Killing goats, pulling thorns.* Grand Rapids, MI: Zondervan.

Tamir, P., & Amir, R. (1981). High school as viewed by college students in Israel. *Studies in Educational Evaluation, 7*(2), 211-225.

Taylor, P. C., Fraser, B. J., & White, L. R. (1994, March). *A classroom environment questionnaire for science educators interested in the constructivist reform of school science.* Paper presented at the annual meeting of the National Association for Research in Science Teaching, Anaheim, California.

Thagard, P. (1992). *Conceptual revolutions.* Princeton, NJ: Princeton University Press.

Torrance, E. P. (1969). *Creativity.* Belmont, CA: Dimensions.

Varrella, G., Kellerman, L., & Penick, J. (1993). In R. E. Yager (Ed.), *Student teaching handbook.* Iowa City, IA: University of Iowa, Science Education Center.

Wheatley, G. H. (1991). Constructivist perspectives on science and mathematics learning. *Science Education, 75*(1), 9-21.

Wiggins, G. (1989a). A true test: Toward more authentic and equitable assessment. *Phi Delta Kappan, 70,* 703-713.

Wiggins, G. (1989b). Teaching to the (Authentic) Test. *Educational Leadership, 46*(7), 41-47.

Wiggins, G. (1993). *Assessing student performance.* San Francisco: Jossey-Bass.

Wiggins, G. (1998). *Educative assessment: Designing assessments to inform and improve student performance.* San Francisco: Jossey-Bass.

Yager, R. E., & McCormack, A. J. (1989). Assessing teaching/learning successes on multiple domains of science and science education. *Science Education, 73*(1), 45-58.

Yager, R. E., & Roy, R. (1993). STS: Most pervasive and most radical of reform approaches to "Science" education. In R. E. Yager (Ed.), *What research says to the science teacher: The science, technology, society movement* (pp. 7-13). Arlington, VA: National Science Teachers Association.

Index

CORWIN
PRESS

The Corwin Press logo—a raven striding across an open book—represents the happy union of courage and learning. We are a professional-level publisher of books and journals for K–12 educators, and we are committed to creating and providing resources that embody these qualities. Corwin's motto is "Success for All Learners."